Springer Biographies

The books published in the Springer Biographies tell of the life and work of scholars, innovators, and pioneers in all fields of learning and throughout the ages. Prominent scientists and philosophers will feature, but so too will lesser known personalities whose significant contributions deserve greater recognition and whose remarkable life stories will stir and motivate readers. Authored by historians and other academic writers, the volumes describe and analyse the main achievements of their subjects in manner accessible to nonspecialists, interweaving these with salient aspects of the protagonists' personal lives. Autobiographies and memoirs also fall into the scope of the series.

More information about this series at http://www.springer.com/series/13617

Hideo Mohri

Imperial Biologists

The Imperial Family of Japan and Their
Contributions to Biological Research

 Springer

Hideo Mohri
University of Tokyo
Bunkyo-ku, Tokyo, Japan

Translated by
Yoko Kawazoe
Tsukuba, Ibaraki, Japan

Editorial advisor
Naoto Ueno
National Institute for Basic Biology
Okazaki, Aichi, Japan

ISSN 2365-0613   ISSN 2365-0621   (electronic)
Springer Biographies
ISBN 978-981-13-6755-7   ISBN 978-981-13-6756-4   (eBook)
https://doi.org/10.1007/978-981-13-6756-4

Library of Congress Control Number: 2019932827

Translation from the Japanese language edition: *Tennoke to Seibutsugaku* by Hideo Mohri © 2015 Hideo
Mohri, published by Asahi Shimbun Publications Inc. All Rights Reserved.

Cover illustration: *Front cover*—Emperor Akihito engaging in research on gobioid fish in 1984
(Courtesy of the Imperial Household Agency).
Back cover—Above: Emperor Showa examining a specimen of sea urchin at the Shimoda Imperial Villa
in 1985 (Courtesy of the Imperial Household Agency); Below: Prince Akishino in front of the panel of
fowls at SOKENDAI (The Graduate University for Advanced Studies) (Courtesy of Akifumi Oikawa)

This Springer imprint is published by the registered company Springer Nature Singapore Pte Ltd.
The registered company address is: 152 Beach Road, #21-01/04 Gateway East, Singapore 189721,
Singapore

Imaginary Family Tree

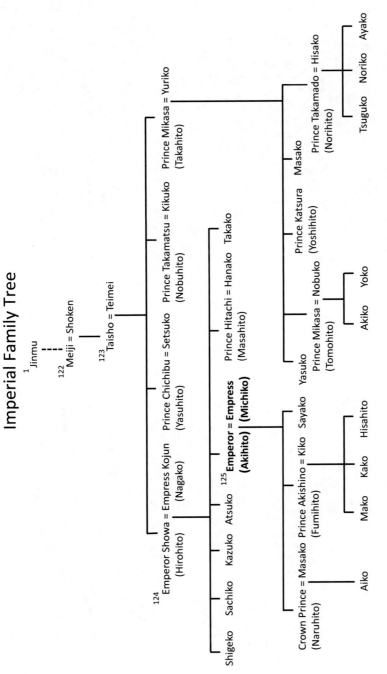

Imperial Family Tree

Former Emperors and Empresses are called by posthumous names in Japan

Emperor Hirohito (Showa) and his family in 1972. Upper left to lower right. Crown Princess Michiko, Corwn Pince Akihito, Emperor, Empress Nagako (Kojun), Prince Hitachi, Princess Hanako, Princess Yori Sayako, Prince Hiro Naruhito, and Prince Aya Fumihito. *Source* Courtesy of the Imperial Household Agency

Emperor Akihito and his family in 2016. Left to right. Crown Princess Masako, Princess Mako, Crown Prince Naruhito, Princess Aiko, Emperor, Empress Michiko, Prince Hisahito, Prince Akishino, Princess Kako, Princess Kiko. *Source* Courtesy of the Imperial Household Agency

Preface to the English Edition

It was decided that Emperor Akihito would abdicate the throne on April 30, 2019. He had expressed his intention to step down primarily as a result of difficulties in fulfilling official duties at his advanced age. Emperor Akihito's sincere and earnest commitment to his role can also be seen in his work as a biologist in his taxonomical researches on gobies. His father, the late Emperor Showa (Hirohito), also conducted outstanding research as a biologist in the taxonomy of marine invertebrates, including hydrozoans and slime molds (myxomycetes), even during those difficult years before and after World War II. Prince Hitachi, the younger brother of Emperor Akihito, is an expert in comparative oncology, Prince Akishino, his second son, is carrying out phylogenetical researches on catfish and fowl and research on biosophia studies, and Sayako Kuroda, his first daughter, is researching birds, including the kingfisher. It is highly unusual that members of three generations of one monarchical family line have devoted themselves to biological studies. We are now hoping that His Majesty the Emperor will be able to enjoy his biological researches to the fullest after his retirement as Emperor.

The fact that the Emperor is a biologist is only vaguely known, even to the Japanese people. When I, as a biologist, prepared to publish *Tennoke to seibutsugaku* (The Imperial Family and Biology, Asahi Shimbun Publications, 2015), I realized that there were only a few works that depicted the Imperial Family members as biologists. With such limited resources, it is even harder for non-Japanese people to obtain a clear understanding of the Imperial biologists of Japan. Given these circumstances, I decided to translate this original Japanese book into English in the hopes that it will introduce the sincere contributions of members of the Japanese Imperial Family to the field of biology to the world. This book also devotes some pages to the International Prize for Biology, which was set up to commemorate the contributions of the two generations of Emperors to biology.

The cover image of the Japanese version, *Tennoke to seibut-sugaku* (The Imperial Family and Biology, Asahi Shimbun Publications, 2015)

I am pleased to have had the good fortune to have the English version published by Springer, thanks to Ms. Yoko Kawazoe, an excellent translator. Several years have passed since the Japanese version was published; therefore, some additional material is included. For readers' reference, the family tree of the Imperial Family and two photographs depicting members of the Imperial Family are newly added at the beginning. The content and wording have been put forward in as simple a manner as possible for readers who are unfamiliar with Japanese history and culture.

I express my thanks to the many persons concerned, including Prof. Naoto Ueno of the National Institute for Basic Biology, Prof. Tetsuji Nakabo of Kyoto University, and Dr. Hiroshi Namikawa of the National Museum of Nature and Science for their help with the translation. I also express my deepest thanks to Mses. Alexandrine Cheronet, Aiko Hiraguchi, Fumiko Yamaguchi and Momoko Asawa of Springer, and Yumiko Nara of Asahi Shimbun Publications for their support in publication.

Tokyo, Japan   Hideo Mohri

Preface

In November 2011, Emperor Akihito was hospitalized for two weeks at The University of Tokyo Hospital for treatment for mycoplasma pneumonia. Deeply concerned by the disaster of the Great East Japan Earthquake of March 11, 2011, Emperor Akihito and Empress Michiko had continued visiting the quake victims and afflicted areas almost every week, in addition to their already busy schedule of official duties. One can imagine how tough this must have been, and these events must have caused fatigue to build up in the Emperor. In February of the following year, Emperor Akihito had to undergo heart bypass surgery. On March 11, 2012, still not fully recovered after the surgery, he attended the memorial service to commemorate the first anniversary of the Great East Japan Earthquake. The following May, he visited the UK to celebrate the Diamond Jubilee of Her Majesty Queen Elizabeth II.

The Emperor and the Empress were scheduled to attend the award ceremony of the 27th International Prize for Biology, to be held at the Japan Academy in Ueno, Tokyo, on November 28, 2011, as one of their official duties. As described in detail later, this prize was founded in 1985 to commemorate the 60-year reign of Emperor Showa (Hirohito) and his longtime devotion to biology and to promote biological researches. Moreover, on the occasion of its 25th anniversary in 2009, it was decided that the prize would also honor the path toward the further advancement of biology, including in commemoration of the achievements of the Emperor at that time, His Majesty Emperor Akihito, who had strived over many years to advance the study of the taxonomy of gobioid fishes while contributing continuously to the development of the prize. Emperor Akihito was discharged from the hospital only two days before the award ceremony. Presumably, his early discharge was because he intended to attend the ceremony. He looked forward to attending it every year as one of the biologists. Unfortunately, his wish did not come true, as he had still not fully recovered his health, and the ceremony was conducted with the hastily arranged attendance of the Crown Prince Naruhito on behalf of Emperor Akihito. Surprisingly, however, the Empress was present at the buffet party celebrating the prize winner, as she had been in past years. The Empress told me that she had made a special point of attending the party on behalf of the Emperor, who was very regretful about not being able to be there that year himself.

These episodes regarding the former Emperor Showa and the present Emperor Akihito show their deep commitment to biology. Moreover, it is well known that not only has Prince Akishino been studying catfish and domesticated chicken, but Prince Hitachi and Sayako Kuroda (the former Princess Nori Sayako) are also engaged in biology-related works. Since Prince Hisahito, the first son of Prince Akishino, appears to be interested in insects and other various living creatures, it would be expected that four generations of the Imperial Family members, starting with the first-generation Emperor Showa, will somehow be involved in biology. This is indeed very rare and unlikely to be seen in any other family of royals in the world. Fortunately, the following year, the Emperor Akihito showed up in much better health at the award ceremony for the International Prize for Biology.

In my boyhood days, Emperor Showa, as the living god, and also as the Generalissimo of the Imperial Japanese Army and Navy at that time, reviewed the Army's military parade, sitting astride his favorite horse "Shirayuki" (meaning "White Snow"), and the Navy's fleet, boarding the battleship "Hiei," an Imperial flagship. A special frame, called a "Hoanden," was provided at every school to house an Imperial portrait of the Emperor and the Empress, and all schoolboys and schoolgirls would offer a profound bow every morning in front of the Hoanden. After World War II, the Emperor's role changed and, according to the new Japanese Constitution, he became a symbol of the unity of the Japanese people, instead of a living god. The Emperor himself issued the proclamation of the "Imperial Declaration of Humanity," which was a public denial of his deity, and commenced tours across Japan so as to meet his people. After the War, I entered the Department of Zoology, Faculty of Science, University of Tokyo. In winter, while participating in a practical course at the Misaki Marine Biological Station, University of Tokyo, as a sophomore, I, together with my classmates, had the honor of seeing the Emperor and Empress (Kojun), who were visiting the station to conduct the Emperor's fieldwork in investigating marine organisms. When we saw him off at the pier, His Imperial Majesty nodded at us from the collecting ship the "Hayama-maru." We did indeed have the honor of receiving gratitude from His Imperial Majesty. It was the one and only time in my life that I met Emperor Showa in person. However, I occasionally had a chance to hear about Emperor Showa from Ichiro Tomiyama, the director of the Station, who used to be his teacher at the old education system high school and his superior after graduation, because Tomiyama helped with the Emperor's research for many years at the Biological Laboratory of the Imperial Household. Also, Shigeko Higashikuni (the former Princess Teru Shigeko), the eldest daughter of Emperor Showa, had been to the station.

I have had the honor of seeing Emperor Akihito and the Empress Michiko every year for the duties of the above-mentioned International Prize for Biology and have had occasions to visit the Imperial Palace to give lectures before them concerning the achievements of recipients of the prize. On such occasions, I happened to see Sayako Kuroda, the then Princess Nori. Prince Hitachi, when he was still Prince

Yoshi, sometimes spent time at the Department of Zoology, Faculty of Science, University of Tokyo, to conduct his research, during which I had conversations with him. I went to see Prince Akishino to ask a favor of him that he might assume the honorary presidency of the International Biology Olympiad, held in Japan in 2009; the Prince kindly accepted this offer, also agreeing to give the opening remarks. Furthermore, when I was a child, I lived in Chiba Prefecture and used to go to the Goryo (Royal) Farm in Sanrizuka where my uncle worked. Looking back over the years, I have had occasional instances of contact with the members of the Imperial Family, despite my status as a mere ordinary person.

In the old days, when people asked what I was studying, the typical reaction to my responding "biology" (or "zoology") was, "Oh, the same as what the Emperor is studying." In the early days after the end of World War II, studying the same subject as the Emperor could have been seen as something noble. Yet, there was probably more sarcasm in those reactions than admiration, as if they were saying "You are in an enviable position, because you can study what you want!" or "What's the point of engaging in such unproductive subjects?"

Biology originated in Ancient Greece and was developed through both Darwin's Theory of Evolution from the middle of the nineteenth century and the discovery of the DNA double helix by Watson and Crick roughly a century later. It has advanced even further as a result of the complete sequencing of the human genome at the beginning of this century, by which the functions of substantially all genes and their regulatory factors are being elucidated, not only in humans, but in other organisms as well. In 2012, Shinya Yamanaka won the Nobel Prize in Physiology or Medicine for iPS cells, the second Japanese winner following Susumu Tonegawa, who elucidated the mechanism of the diversity of antibody genes. Furthermore, Satoshi Omura and Yosinori Osumi won the prize for the discovery of pharmaceuticals originating from microorganisms and the discovery of autophagy, respectively (further Tasuku Honjo has got the prize for the establishment of a new type of cancer therapy "immunotherapy"). Superficially, a new era of "bio" is blooming. Unfortunately, however, biology, or the "life sciences," is not sufficiently understood in our country, even today. I must say that neither the industrial sector nor the government have been willing to allocate sufficient budgets to biological researches.

This book describes how the members of the Imperial Family, including Emperor Showa, became interested in biology, their achievements, and how those achievements have been appreciated by experts in their respective fields. Many episodes concerning Emperor Showa are told from a variety of viewpoints, with most of them springing from political and social aspects related to the turbulent Showa period (1926–1989). Even in regard to the present Emperor, when he visited the UK in 1998, the media focused mostly on the protest by the former English captives, and few media mentioned the fact that his research on fish had been very well received and that he was given a special medal (the Charles II Medal) by the Royal Society for said achievement.

In this book, focusing exclusively on the involvement of the members of the Imperial Family in biology, it is my intention to describe their backgrounds and actual activities, and the way in which their achievements have been appreciated, both at home and abroad. The term "biology," as used in this book, means biology in terms of academic disciplines: mathematics, physics, chemistry, and geology. It is also called "life science," in contrast to the term "material science," in Western countries and also in Japan in some cases. However, in Japan, the term "life science" is often used in various areas of application, including the clinical application of medical science. Therefore, it is sometimes called "biological science," for which there is no corresponding word in Western countries. In those countries, the term "life" means not only the anima, but also general livelihood. Focusing on the latter, the work regarding water traffic by Crown Prince Naruhito, which appears to bear no relation to biology, may be very close to life science.

Moreover, most works by the three-generational Imperial Family members, Emperor Showa, Emperor Akihito, and Prince Akishino, could be categorized into an old discipline, "natural history." Some may regard this as an obsolete discipline, because of its long history. Nevertheless, that is not the case. Knowledge of natural history, which consists mainly of taxonomy, as well as ecology, is essential for addressing the issues of biodiversity that are critical for us earthlings to survive well into the future. Furthermore, taxonomy and phylogeny provide the bases for biological evolution, which has attracted a lot of interest. It is necessary that we identify the species of animal and plant specimens used in experiments and determine whether these specimens belong to the same species or to the same lineage. New mammals are still being discovered, creating a sensation. In fact, among all of the living organisms (about 100 million species) on the earth, only a small fraction (about 1.8 million species) has been given respective scientific names to describe them in the register. In this book, I will also consider the situation in which "biology" now finds itself in Japan, in connection with these problems.

Finally, I had no choice but to rely on literature and documents previously published when describing, in particular, Emperor Showa, because all of those who were closely associated with him have passed away. Unfortunately, that is equally applicable to Emperor Akihito and Prince Akishino. Fundamentally, I referred to the diaries written by the former grand chamberlain, Sukemasa Irie, and the former chamberlain, Ryogo Urabe ("Irie Sukemasa's Diary," edited by Tanetoshi Irie, [26] and "Diary of Ryogo Urabe, The Last Chamberlain of Emperor Showa," co-edited by Takashi Mikuriya and Katsumi Iwai [40]), for Emperor Showa, and to the book "A Chamberlain of the Imperial Family—Serving as a Grand Chamberlain for Ten and a Half Years," written by the former grand chamberlain, Makoto Watanabe, [68] for the present Emperor. I also referred to information posted on the Web site of the Imperial Household Agency. The references are listed at the end of the book, because only some of the sources are specified in the text. For information on the publication dates of these references, refer to the list. The titles of persons cited in the text were all omitted. I appreciate their acceptance of my sincerest apologies for these omissions. I am entirely responsible for the wording of the book, including in regard to the matters mentioned above.

The Imperial Household Agency released the contents of "Showa Ten'nou Jitsuroku" ("The Annals of Emperor Showa"), [21] composed of 12,000 total pages, on September 9, 2014. New facts were revealed after I finished writing the chapter concerning Emperor Showa, but I only quoted some of the articles, including a serialized article, "Deep Interpretation of The Annals of Emperor Showa," from Asahi Shimbun at my discretion, because I thought that only some of them were relevant to the image of Emperor Showa as a biologist.

Tokyo, Japan   Hideo Mohri

Contents

Introduction

Royalty and the World of Learning

As history has demonstrated, academics and the arts have long developed under the patronage of those in power at the time. It is well known that Crown Prince Naruhito plays viola and sometimes joins an orchestra as one of its members. Similarly, certain European royals are known to play some variety of instrument in furtherance of their enjoyment of music. However, in many cases, people in their position serve only as patrons of artists, holding concerts, hiring painters to create portraits of them and members of their family, and collecting various works. The same thing can be said of the academic field.

With the dawn of the age of exploration in the fifteenth century, European countries began to make forays into unexplored regions, such as Africa, America, Australia, and East Asia, and expanded their activities globally up through the eighteenth century, leading to colonies being formed all over the world. This effort reached its peak during the years of Queen Victoria in the nineteenth century, and, as a result, England (hereafter, referred to as the UK) reigned over one quarter of the world, not only in terms of its population but also the surface area of the earth.

Accordingly, unusual cultural items and aspects were brought to the suzerain from those colonized countries throughout the world. The trans-global voyage of the H. M. S. Beagle, during which Charles Darwin formed his theory of evolution, is the best example of the above situation. Moreover, plant hunters searched the entire world for rare plants, and a vast number of botanical specimens were collected and archived in the museums of European countries, including the Natural History Museum, London. Zoos and botanical gardens were built in these countries, allowing the Europeans to see exotic animals and plants that they had never seen before. One example of this is Kew Gardens in the UK, which houses a huge number of botanical specimens, estimated at about seven million.

Focusing on our country, Japan, Philipp Franz von Siebold (1796–1866), who came to Nagasaki under the national isolation system, took animal and botanical specimens collected in Japan back to the Leyden Royal Museum of Natural History

(Naturalis). The analytical observation of these specimens ultimately came to fruition, resulting in the publication of "Flora Japonica" and "Fauna Japonica." The royal families and nobles in the countries concerned played a significant role in facilitating these achievements and thus were not shy in taking all of the credit for and showing great pride in them. The nobility system (in Japan, the former court nobles and feudal lords, as well as meritorious retainers in the Meiji Restoration, were appointed as nobles during the period from the Meiji period to the end of World War II) consists of dukes, marquises, counts, viscounts, and barons, in descending order of rank.

The royals and titled nobilities also seemed to be attracted to natural history. Antonie van Leeuwenhoek, who had discovered spermatozoa under his originally developed monocular microscope in 1677, described, in his letter to the Royal Society, how James II of England was quite surprised and laudatory when Robert Hook, a discoverer of cells known for Hook's law, one of our bedrock physical laws, showed him spermatozoa under the microscope. Karl von Linné (1707–1778), from Sweden, known as "the Father of Taxonomy," systematically classified a huge number of animals and plants, as well as minerals, using binomial nomenclature, taking full advantage of the collections owned by King Adolf Frederic of Sweden, Queen Ulrika Eleonora of Sweden, Count Carl Gustav Tessin, and other well-known people of wealth from Europe.

In recent years, one of the royals known to have conducted researches himself is Albert I, Prince of Monaco (1848–1922), a famous marine biologist. He is the great grandfather of Rainier III, Prince of Monaco, well known for his marriage to Grace Kelly, a Hollywood actress from the USA. Albert I constructed a marine research vessel for the purpose of making a systematic survey of marine organisms in the North Atlantic and the Mediterranean from the equator to the Arctic Circle. He racked up many achievements, including discovering a variety of new species of deep-sea creature, and even witnessing a giant squid being eaten by a sperm whale. Moreover, he built a marine museum in Monaco with an aquarium to exhibit many of said specimens, which he himself had collected. In addition, he drew a chart of the currents that flow through the Atlantic Ocean, including the Gulf Stream, greatly contributing to marine physics. He also built and presented a marine research institute in Paris. However, Albert I did not necessarily conduct research or write theses himself, as he put all of the descriptions of the new species into the hands of professionals. Although he accompanied the research vessels, his role in the research itself was generally that of a patron or a sponsor.

The former King Gustav VI Adolf (1882–1973) of Sweden, a grandfather of the present King of Sweden, was also known to be an expert in archaeology and botany, and he had deep knowledge of oriental cultures. He engaged in eager observation at archaeological sites such as Nara and Kyoto upon his visit to Japan in 1926, when he was still Crown Prince, and made a tour of inspection of an evacuating site in Gyeongju, Korea, which was a Japanese colony at that time. It is said that, after returning to Sweden, he publicized his speculation that the potteries exhibited in Shosoin in Nara, about which a theory of tricolored glazed potteries (tángsāncǎi) of the Tang Dynasty prevailed at that time, were made in Japan, namely, that they might be referred to as Nara tricolored potteries. It is not too much to say that he was simply an academically minded king.

I had a chance to attend the Nobel Prize Award Ceremony held at a concert hall situated in Stockholm on December 10, 1963, when I was studying in Sweden. That day was also the anniversary of the death of Albert Nobel, the founder of the Prize. In 1963, Alan Lloyd Hodgkin and Andrew Huxley, of the UK, as well as John Carew Eccles, of Australia, won the Nobel Prize in Physiology or Medicine. Surprisingly, the winners were seated on the stage, and the Swedish Academy members were seated on the upper tier, while the royal family members, including King Gustav VI, had their seats situated below the stage. This would be an unimaginable scenario in Japan (note that, in recent photographs, they sit on seats placed on the stage). This episode shows the degree of importance that is given to academics by the Kingdom of Sweden and the Swedes, including the king.

As described earlier, the members of the Imperial Family line in Japan, Emperor Showa, Emperor Akihito, Prince Hitachi, and Prince Akishino (as well as Sayako Kuroda), are unique characters in the world, in that they are not just patrons, but also, with some exceptions, who leave the work of identification and description of collected specimens to experts, likely to visit the target fields themselves to collect animal or plant specimens, make observations about and carry out experiments on them, and write academic papers on biology. Similarly, Prince Chichibu, known as the "Prince of Sports," and the only one of Emperor Showa's brothers who ever went to the UK to study at the University of Oxford, in 1925 when he was a lieutenant, seemed to be interested in biology. Prince Mikasa (the youngest brother of Emperor Showa) and the Crown Prince Naruhito are professionals in ancient oriental history and water transport, respectively. The works of both men have been held in high regard by foreign academia. I, for one, would like to see the Japanese people take more of an interest and be prouder of their Imperial biologists.

The Imperial Family in Japan and the World of Learning

Turning our eyes to the Imperial Family in Japan, we will take a detailed look at the contributions of the family members to academics and the culture as a whole. In many cases, the successive Emperors have been directly involved in, above all, poetry and music, as well as calligraphy. Their involvements are outlined below by reference to "Rekidai Tenno Souran—Koi wa Do Keisyosaretaka" ("Manual of the Successive Emperors – How the Imperial Throne Has Been Succeeded"), written by Hidehiko Kasahara [34].

The Emperors who are counted among the nation's outstanding poets are numerous. For some examples, the names of eight Emperors out of the successive Emperors from the 38th Emperor Tenji to the 84th Emperor Juntoku appear in "Hyakunin Isshu" ("One Hundred Poets, One Poem Each"), edited by Teika Fujiwara in the Kamakura period (1185–1333). In addition, the 60th Emperor Daigo compiled "Kokin Wakashu" ("Collection of Ancient and Modern Japanese

Poetry"), the first Imperial Waka Anthology, and the 84th Emperor Go-Toba, one of 100 poets selected in "Hyakunin Isshu," helped form the compilation "Shin Kokin Wakashu" ("New Collection of Ancient and Modern Japanese Poetry"). In recent years, the 115th Emperor Sakuramachi, the 117th Empress Go-Sakuramachi, the last female Emperor in Japan, and other Emperors have composed excellent waka poems.

Famous for being a skilled calligrapher, the 52nd Emperor Saga was reputed to be one of the three great calligraphers of the Heian period (794–1185), along with Kukai (a famous Buddhist monk) and Hayanari Tachibana. During his reign, "Ryoun Shu," the first anthology of Chinese poems in Japan, was compiled at his command. In the medieval time, the 104th Emperor Go-Kashiwabara, who was also a skillful calligrapher, transcribed "Han'nya Shingyo" ("Heart Sutra") by himself and dedicated it to Enryaku-ji and Nin'na-ji, two temples in Kyoto, when the smallpox epidemic occurred in the sixteenth century. The next 105th Emperor Go-Nara sold Imperial autographs to shore up the financial conditions of the Imperial Household, which were suffering from the aftereffects of the Onin War (1467–1477). As for music, the 53rd Emperor Jun'na was the first Imperial leader to play the Koto (a 13-stringed Japanese zither), and some Emperors, including the 73rd Emperor Horikawa, were said to be masters of the Koto.

Focusing on academics, it is said that Prince Shotoku (574–622), who attained various achievements as the Crown Prince during the reign of the 33rd Empress Suiko, the female Emperor, compiled such books describing the history of Japan as "Tennoki" and "Kokuki." "Nihongi" ("The Chronicles of Japan") was compiled by Prince Toneri (676–735), a son of the 40th Emperor Tenmu, and others during the reign of the 44th Emperor Gensho. Since then, "Rikkokushi" ("The Six National Histories"), which includes "Shoku Nihongi" ("The Chronicle of Japan Continued") and "Nihon Kouki" ("Later Chronicles of Japan"), describing the history of Japan up to the 58th Emperor Koukou, have been compiled at the commands of successive Emperors. However, the further history of Japan was not compiled until modern times. The 84th Emperor Juntoku, one of the more academically inclined Emperors, mastered the fundamentals and origins of ceremonies and manners passed down in the Imperial Household and wrote "Kinpi Sho," a practical guide to these traditions. The 95th Emperor Hanazono of the Jimyoin Imperial Lineage (Northern Dynasty), who was a voracious reader of literature, both classical and otherwise, was celebrated as the greatest font of knowledge of the day and for his pursuit of the ideal attitude for an Emperor.

Hideyoshi Toyotomi (1537–1597), a daimyo warrior who unified the country during the Sengoku period (1467 (or 1493)–1590), treated the Imperial Family members extremely politely, as seen in the instance when the 107th Emperor Go-Yozei visited his residence, known as Jurakudai, but the subsequent ruler, Ieyasu Tokugawa (1543–1616), made efforts to maintain the predominance of the Tokugawa Shogunate over the Imperial Court. The establishment of "Kinchu Narabini Kuge Shohatto" ("Code for the Emperors and Nobles") in 1615 obviously reflects his attitude, which was marked by his feeling of superiority in power to the Imperial Court. In this way, the Imperial Court remained deprived of its political sovereignty by the Shogunate until the Restoration of Imperial Rule in 1867. Against this background, most of the Emperors in the Edo period (1603–1867) became dedicated to learning. For example, the 108th Emperor Go-Mizunoo invited Ingen Zenji (posthumous name), a Buddhist from China, to hold study sessions, and he also acquired deep knowledge of foreign affairs. The 110th Emperor Go-Komyo, the 111th Emperor Go-Sai, and the 112th Emperor Reigen, all sons of Go-Mizunoo, were also fond of learning, as seen in their respective devotion to Confucianism and maintenance of the recording office, which prepared and organized duplicates of the records of the Imperial Court.

Like the above-listed Emperors, the 119th Emperor Kokaku, the 120th Emperor Ninko, and the 121st Emperor Komei (reign from the late Edo period to the end of the Edo period) were also devoted to study, and Emperor Ninko established the foundation of the current Gakushuin University. The 122nd Emperor Meiji (1852–1912) vigorously studied academics and morality under the gracious tutelage of Nagazane Motoda, Hiroyuki Kato, and others, and the 123rd Emperor Taisho (1876–1926), who had an interest in Chinese classical literature, composed many Chinese poems, unlike Emperor Meiji and the 124th Emperor Showa (Hirohito, 1901–1989), who were fond of waka poems. Emperor Taisho was also interested in biology. Despite being in a politically delicate position, indeed, often because of that very position, the Emperors were able to devote themselves to academics.

Looking back on the history of the Imperial Household, it is not by chance that the Emperor Showa became interested in biology and history. The members of the current Imperial Family have inherited an academically oriented nature from generation to generation, especially from the Emperors in modern times. I have frequently heard the saying, "A third generation savant." This means that, even in an academic household, it may take three generations to produce a successful scholar. There is no doubt that Emperor Showa built the foundation for three generations of biologists. However, the preceding Emperors were also academically minded Emperors. Emperor Akihito stated, at a press conference held in the summer of 1977:

"'The pen (scholarship)' rather than 'the Sword (martial arts)' was traditionally passed down in the Imperial Household from generation to generation. Only a few Emperors were dressed in military uniform. I intend to preserve that Imperial Tradition that is fond of 'the Pen.'"

Chapter 1
His Majesty Emperor Showa (Hirohito) [1901–1989]—The First of Three Generations of Imperial Biologists

The Personal History of Emperor Showa

Emperor Showa was born to the then-Crown Prince Haru-no-Miya Yoshihito (later Emperor Taisho) and the then-Crown Princess Sadako (later Empress Teimei) as the eldest son of the Crown Prince's Palace on April 29th, 1901, and was named Michi-no-Miya (Prince Michi) Hirohito (hereafter, simply referred to as Hirohito). According to a well-informed source, since his father, the then-Crown Prince, was weak when he was born, the birth of a healthy baby greatly delighted his grandparents, Emperor Meiji and Empress Dowager Shoken. In the following year, Prince Atsu Yasuhito (later Prince Chichibu) was born as the second son. The two infant Princes were fostered out to the family of Earl Sumiyoshi Kawamura, the former Lord of the Admiralty and the then Privy Councillor, in accordance with the traditional court customs, with great prayers for the healthy growth of the children born into the Imperial Family. Regrettably, Earl Kawamura passed away in 1904, and the Princes moved to the Koson Goten (the Imperial Grandchildren's Palace), which was next door to the Crown Prince's Palace, connected through a garden, and lived a life with their parents, the Crown Prince and the Crown Princess, in a lifestyle almost the same as that of ordinary people. In the following year, 1905, Prince Teru Nobuhito (later Prince Takamatsu) was born as the third son, and soon joined his brothers, all three Princes then being brought up in the Koson Goten. It is said that Emperor Taisho often dropped into the Koson Goten with Empress Teimei, and sometimes enjoyed playing tag with his sons and their fellow schoolboys (Osanaga Kanroji, "Sebiro-no-Ten'nou" ("The Emperor in a Suit") [33] (Fig. 1.1).

"The Annals of Emperor Showa" [21] exclusively describes scenes in which Prince Hirohito was not free to see his father, Emperor Taisho, though the description does not seem to be based on fact. Prince Sumi Takahito (later Prince Mikasa), who was 14 years younger than Emperor Showa, was born as the fourth son in 1915 and was fostered in the Imperial Palace, unlike his three elder brothers.

© Springer Nature Singapore Pte Ltd. 2019
H. Mohri, *Imperial Biologists*, Springer Biographies,
https://doi.org/10.1007/978-981-13-6756-4_1

Fig. 1.1 Emperor Showa, Hirohito, in his childhood (second from the right), together with his father, Emperor Taisho, when he was the Crown Prince (on the right side), and his younger brother, Prince Chichibu (second from the left) (Courtesy of the Imperial Household Agency)

Emperor Meiji was greatly concerned about the physical condition of his grandchildren, because his Crown Prince (later Emperor Taisho) had suffered from a severe illness when he was a child, and the Emperor often instructed them to go to the Imperial Villa at Numazu, Atami or Hayama to rest in order to adapt themselves to a physically active lifestyle. In fact, in his childhood, Emperor Showa often enjoyed watching sumo-wrestling. He wrestled himself with his fellow schoolboys, and swam very often in Sagami Bay, which faces Hayama-Machi, where the Hayama Imperial Villa is situated. He collected shells at the Numazu and Hayama Imperial Villas and also explored places suitable for insect and plant collection. Using the Prince's admission to the Gakushuin Primary School as an opportunity, Emperor Meiji appointed General Maresuke Nogi, who was a serious and honest man, to be the president of the Gakushuin. Emperor Showa built his serious, scrupulous and fair personality under the tutelage of President Nogi through learning at the Gakushuin School and the President's creed that "one is required to make best efforts to discipline oneself." Such a quality is essential to the personality of researchers.

Emperor Showa had long been interested in living organisms, especially animals, starting in his childhood, as evidenced by the fact that he visited the Ueno Zoo 14 times in total, with a frequency of about two times every year, during the period from four to ten years old ("The Emperor's Biological Research," National Museum of Nature and Science) [63]. He not only closely watched the animals being raised there, but also listened hard to the explanatory words of the director of the zoo. According to "The Annals of Emperor Showa," [21] he also visited the zoo several times after then. Kanroji introduced, in his book, an anecdote related to Emperor Showa, who has always been know to treat animals with humanity:

> When the Emperor was a child, a raccoon dog was captured on the Imperial estate, the current Meiji Jingu, a Shinto shrine where Emperor Meiji and Empress Dowager Shoken are enshrined. The attendants were struck by the idea that the raccoon dog should be shown to the Imperial Prince, Hirohito, a grandchild of Emperor Meiji. However, there was a hound-breeding farm on the Imperial estate, and when the captured raccoon dog was led before the Imperial Prince Hirohito, the dogs there all bayed at it in unison. The raccoon dog was all atremble with fear. Prince Hirohito, when saw the scene, said,
>
> "That is enough. I do not have to see the animal. Put it back in animal housing right now, please,"
>
> And he returned to the Crown Prince's Palace. Afterwards, he repeatedly asked, "What has the raccoon dog been doing?"
>
> It was quite clear that his concern for the animal's well-being was both genuine and pervasive [33].

In regard to insects, he seemed to learn the names of individual species by referring to a magazine, "The Insect World," and was fond of reading "One Thousand Insects of Japan" by Shonen Matsumura, Japan's first illustrated guide to Japanese insects. He expressed a strong interest in the scene at the Sericultural Institute, where researches on silkworms and raw-silk threads, as well as the activities regarding public awareness of them, were conducted. After his first time there, he expressed a great hope to be able to visit again, saying it was "more fun than the zoo."

At a press conference held at the Nasu Imperial Villa in August 1965, Emperor Showa told a heartwarming story:

> I remember that I climbed a mountain (Mt. Haruna) near Ikaho twenty-one times for insect collection, during which I caught two great purple emperors (*Sasakia charonda*) and made a lantern procession with my brother, Prince Chichibu, to celebrate the catch. The great purple emperor, Japan's national butterfly, is a large, beautifully-colored butterfly belonging to the family Nymphalidae, and any boy familiar with insects eagerly desires to catch it.

From the story about how Prince Hirohito climbed the mountain many times until he was finally able to catch the butterfly, it can be easily imagined just how happy he was with the catch. However, this story notwithstanding, the Prince, who had a kind, considerate personality, provided an answer to the question posed by Edred John Henry Corner, a botanist and the first recipient of the International Prize for Biology, "What led Prince Hirohito to the study of biology?": "I became interested in biology for the first time when I made a research on wild plants in Shiobara at the age of 12 (Edred John Henry Corner, "His Majesty Emperor Hirohito of Japan, K.G." [9]).

The Prince first visited the Misaki Marine Biological Station, University of Tokyo (Imperial University of Tokyo, 1886–1947), with Prince Yasuhito at the age of 9 on September 6th, 1910, the year before the Station opened up the aquariums to the public.

During their visit to the Station, an unexpected event occurred. At the very moment that Prince Hirohito scooped a squid up out of the aquarium water with a "fishing net," the squid ejected its ink onto Kumakichi Aoki (usually known by his nickname, Kuma-san), the collecting fisherman at the Station, who was standing next the Prince, causing his white formal wear to get stained with the ink. Seeing that scene, the Prince said:

Aoki, don't be angry, because I will send you a new outfit after I return to the Palace.

Although the Prince later visited the Station several times, he never brought up this matter again. As a result, Kuma-san proudly stated,

Throughout all of Japan, I am the only one to whom the Emperor owes something.

The Prince heard about this from Masamitsu Oshima, who was involved at the Misaki Marine Biological Station at that time, through his chamberlain. Therefore, when visiting the Station in 1929, he granted Kuma-san money through his chamberlain, saying,

My debt is now paid.

Thus was this particular source of pride taken away from Kuma-san (Masamitsu Oshima [52]).

By the way, Kuma-san was a popular collector who worked at the Station from the Meiji and Showa Periods through to the Taisho Period. He was highly esteemed by foreign researchers for his ability to orally recite the scientific names of marine animals in Latin. It should be noted that, according to "Showa Ten'nou Jitsuroku," [21] after returning to the Hayama Villa, Prince Hirohito said, "Among all of the sites I have ever visited, the Misaki Marine Biological Station is now my favorite one."

Meanwhile, the Prince was not only interested in animals, but in plants as well, including when he was a child, as we know from his afore-mentioned answer to the question by Edred Corner. During his summer vacation after the sixth grade, he prepared five displays entitled "Insects and Plants" a sort of exhibit of "ecological specimens," in which selected plants were laid out, with the insects associated with these plants arranged around them (Fig. 1.2). It is exceptionally unusual for an insect collector to conceive of such an idea. They tend to pride themselves on their specimens, and their displays are uniformly made up of different examples of the same species of insect, for example, butterflies or stag beetles, simply arranged in a specimen case. Plant collectors are not much different. The specimens in this exhibit showed that the Prince had already acquired the attitude of a naturalist, characterized by the trait of understanding nature as a whole. These specimens were subsequently stored in the specimen room at the Laboratory and are now on exhibition at the Emperor Showa Memorial Museum in Showa Kinen Park, Tachikawa, Tokyo. Moreover, in the same year, the Prince visited Kyoto's Hirase Conchological

Fig. 1.2 "Specimens of
Insects and Plants" collected
by Emperor Showa. Around
two stems of *Patrinia villosa*,
seven specimens of butterfly,
namely, *Pieris melete*,
Celastrina argiolus, *Lycaena
phlaeas*, *Minois dryas*, *Neptis
sappho*, *Lasiommata
deidamia* and *Dichorragia
nesimachus are arranged*
(Courtesy of the Imperial
Household Agency)

Museum (a shellfish museum), built by Yoichiro Hirase, who made a significant
contribution to the conchological field in Japan. This motivated the Prince to become
interested in shellfish and engaged in collecting shellfish so as to examine the names
and characteristics of those he collected with the help of illustrated guides.

Two descendants of royalty, Prince Kacho Hirotada and Prince Kuni Kunihisa,
and ten descendants of the nobility were selected as classmates of the Prince at the
Gakushuin Primary School. Among this group were included Torahiko Osako
(eventually to be renamed Torahiko Nagazumi), who later served the Emperor
Showa as chamberlain and the chief ritualist of the Imperial Court for 80 years, and
Tadao Kuroda (eventually to be renamed Tadao Sato), a developmental biologist,
who later graduated from the Department of Zoology (Zoological Institute), Faculty
of Science, University of Tokyo, and worked as a part-time researcher at the
Biological Laboratory of the Imperial Household and as a professor at Nagoya
University. During the Prince's time at the Gakushuin Primary School, Emperor
Meiji, his grandfather, passed away, and Maresuke Nogi, the president of the
Gakushuin, committed suicide with his sword. When the then Crown Prince (later
Emperor Taisho) acceded to the throne, Prince Hirohito became the Crown Prince
and was appointed as a second lieutenant and ensign.

The Young Crown Prince Enters the World of Learning

After graduating from the Gakushuin Primary School in 1914, Crown Prince Hirohito received a special education in furtherance of becoming Emperor with five fellow students, including Nagazumi, at the Crown Prince's Imperial 'Study,' built within the site of the Temporary Prince's Palace in Takanawa. The philosophies enacted in regard to education, which are considered to have originally been worked out by Nogi, were gradually diminished, changing from Nogi's militaristic style to one that was more unbiased after Nogi's death. At that time, the president was Heihachiro Togo, who was an admiral of the fleet; the vice president was Arata Hamao, who was the lord steward to the Imperial Prince and former president of the University of Tokyo; and the councilors of the School were Naoharu Osako, the Full Admiral, who was an uncle of Nagazumi and the president of the Gakushuin (a successor to Nogi), and Kenjiro Yamakawa, who was president of the University of Tokyo. Reaching into the ranks of the Imperial Court, the academic community and the army and navy, the greatest men of the age were asked to serve as members of the faculty. The enrollment period was seven years, equivalent to the integrated period of middle and high schools in the old education system that was in force before the current system, providing six years of elementary school, three years of middle school, and three years of high school.

According to the book written by Torahiko Nagazumi, "Serving the Emperor for 80 years—Emperor Showa and Me," [47] the Temporary Crown Prince's Palace, to which Crown Prince Hirohito moved so as to live apart from his brothers, was constituted of a large Japanese-style two-story building, a western-style building, a building with classrooms and an adjoining staff room, and a gymnasium. It is said that opening and closing ceremonies were held in the western-style building. All of the Japanese-style buildings (the throne for Hirohito) and the other three buildings were collectively called the "Togu-Gogakumonjo" ("the Crown Prince's Imperial 'Study'"). Also, a separate western-style specimen museum was built in which objects were to be displayed, including offerings such as a penguin gifted to Hirohito by Nobu Shirase, a lieutenant who was well known for making an Antarctic expedition, and the animal and plant specimens collected by Hirohito himself. He often visited the museum to examine the specimens. Besides these specimens, dogs and various kinds of bird were kept in a kennel and an aviary, respectively; furthermore, monkeys had been reared in the area for quite some time. However, none of these animals were the subject of the Crown Prince's study at that time. Later, in 1923, the throne room and the specimen museum were destroyed by fire when the Great Kanto Earthquake hit the Tokyo metropolitan area.

In the Crown Prince's Imperial 'Study,' he learned about a wide variety of subjects, including ethics, under the tutelage of Jugo Sugiura; he studied history, geography, physiography, Japanese language/Chinese classics, natural history, physics and chemistry, mathematics, French, law/economics, art history, calligraphy, gymnastics/martial arts, equestrian art, and military affairs. Kurakichi Shiratori taught Crown Prince Hirohito historical science, the subject in which he was most interested besides biology. According to the book by Nagazumi [47], Shiratori gained the prior consent of Nogi in regard to his basic educational attitude: "I hope to enlighten the Prince and his fellow students, especially about the point that historical facts should be recognized as being independent of myths." Accordingly, Shiratori must have delivered lectures to the Prince on historical science different from those which we, the present author and others, learned in history class, having started from the myths that sprang up during World War II. The Crown Prince was impressed by "A Lecture on Western World History" and "The History of the Great French Revolution," written by Genpachi Mitsukuri, and read them repeatedly after school hours and during his free time. He was very interested in the great persons of history.

Genpachi Mitsukuri, a professor of history at the University of Tokyo, was a grandchild of Genpo Mitsukuri—a doctor who contributed to the modernization of Japan during the period ranging from the end of the Edo Period to the Meiji Restoration, and a member of a well-known family that turned out a number of scholars, as will be described later. Originally, he had graduated from the Department of Zoology, Faculty of Science, University of Tokyo, and then traveled to Germany to study further. However, he suffered from severe myopia, and was therefore unable to manipulate a microscope skillfully, eventually being compelled to convert his major to Western history. The historical literature at the 'Study' was selected by Sadasuke Makino, a chamberlain, at his discretion, according to his own specialty. They say that Heihachiro Togo, the president of the Crown Prince's Imperial 'Study,' and Kinmochi Saionji, a Genro (senator) at that time, had a strong impact on the decision of Emperor Showa to select biology as his research subject rather than history. This is probably because they were concerned about any mental conflict that might arise in the Crown Prince, a future emperor, possibly caused by such problems as the conflict between the Northern and Southern Dynasties (Nanboku-cho) in Japanese history and the revolutions in Western history. Naganari Ogasawara, the vice admiral who was involved in both the conceptualization and realization of the 'Study,' wrote, in his diary, that, contrary to the expectations of those around him, the Crown Prince had a preference for learning natural history over all other subjects, including military affairs. It should be noted that Genros (of whom there were nine in total), previously called Genkuns (veterans), were appointed by order of Emperor Meiji from among those who contributed greatly to the Meiji Restoration. They were the pillars of the state, who were involved at the highest level of national decision-making regarding administrative matters, such as the recommendation of a successor to the prime minister and whether to wage war or make peace with other countries. During the period from 1924 to 1940, the only person who served the state as a senator was Saionji.

Hirotaro Hattori (1875–1965), a botanist, taught Crown Prince Hirohito natural history at the 'Study' (Fig. 1.3). He enlightened the Crown Prince, who was fond of living organisms, as to the way to acquire the culture and basic knowledge essential to biology, which bore fruit, resulting in the Crown Prince growing up to be a world-class taxonomist. Hattori served Emperor Showa as a chief at the Biological Laboratory of the Imperial Household for a long time. He was born into a vassal of the Owari Tokugawa clan at Surugadai, Tokyo. A mycologist, he leaned under the tutelage of Manabu Miyoshi, who coined the term "Seitaigaku" in Japanese from the term "ecology" in English, at the Department of Botany (Botanical Institute), Faculty of Science, University of Tokyo. After graduation, he worked at the University of Tokyo as a lecturer, then at Gakushuin University as a professor, and, finally, was appointed as a chamberlain of the Crown Prince's Imperial 'Study.' Light-hearted and in no way eager for fame, he enjoyed the deep confidence of Emperor Showa. According to Hiroko Daba, a painter working at the Biological Laboratory of the Imperial Household, he was a tall, thin, upright man, as well as being extraordinarily kind (Corner [9]). He had close ties with Kurakichi Shiratori, who was his colleague. Hattori will appear again later in various important scenes in this book.

And then there was chamberlain Masanao Tsuchiya, one of the people responsible for bringing up Emperor Showa during his childhood. Although Tsuchiya graduated with a degree in the Humanities from the University of Tokyo, he was also interested in conchology and gave Emperor Showa behind-the-scenes support in becoming a biologist by making plans for collecting shellfish, insects and plants, attending the Crown Prince during his collection of them, giving guidance

Fig. 1.3 Hirotaro Hattori, a botanist, who supported Emperor Showa's research from his time being educated at the Crown Prince's Imperial 'Study' (Courtesy of Hiromi Hattori)

about how to display specimens and preparing the specimen labels (Kanroji [33]). In addition, Tsuchiya was a fellow student of Emperor Taisho and served at the 'Study' as a lecturer in French.

In the early morning of the day following a storm in March, 1918, the 16-year-old Crown Prince Hirohito found a large red prawn 18.3 cm in length, a size he had never seen before, on the edge of the water near the Numazu Villa (the present Numazu Imperial Villa Memorial Park). The Crown Prince consulted various reference materials but failed to identify the prawn. Therefore, taxonomic evaluation of this prawn was entrusted to Arata Terao, a biologist. Terao, who graduated from the Department of Zoology, University of Tokyo, worked as a professor at the Training Institute of Fisheries (the former Tokyo University of Fisheries and the present Tokyo University of Marine Science and Technology, after incorporation of the Tokyo University of Mercantile Marine). Four years later, he recognized it as a new species and named it *Sympasiphaea imperialis*. Needless to say, the Latin sub-class name *imperialis* means "Imperial" (Fig. 1.4). The prawn had been considered for a long time to be the first new species found by Emperor

Fig. 1.4 Sample of *Sympasiphaea imperials* collected by Emperor Showa in Sagami Bay (Courtesy of the National Museum of Nature and Science)

Showa; regrettably, a joint research by Prince Hitachi, Masahito, the second son of Emperor Showa, and Masatsune Takeda, a scientist at the National Museum of Nature and Science (the former National Science Museum), revealed that this prawn was of the same species as one that had been found in the Atlantic long before the discovery by the then-Crown Prince and had already been registered as *Glyphus marsupialis*. Since then, the nomenclature named after Emperor Showa has not been used for this prawn (Memoirs of the National Science Museum, Vol. 15, 1982). Nevertheless, we still wonder at the considerable size of the prawn found by the Crown Prince, compared with the normal size of about 8 cm. Having been moved to the Showa Memorial Institute, the National Museum of Nature and Science, this prawn specimen is now displayed there, together with a large number of specimens that were once contained in the Biological Laboratory of the Imperial Household. Needless to say, in the time that followed, Emperor Hirohito found many new species of animal and plant, one after another, with the final count being well over one hundred.

There is nothing more exciting for taxonomists and collectors than finding a new species of animal or plant that has not yet been registered. According to Jun'ichi Aoki, a professor emeritus at Yokohama National University, one of the so-called Tonosama Gakusya (noble biologists) described later and discoverer of about 450 new species of the suborder Cryptostigmata (a sort of mite) with a size of about 0.5 mm within and outside of Japan, "It is exciting and gratifying for many taxonomists to find, name and describe new species." [2]. Even regarding known animals, I spend a good deal of time seeking them out, and I can always feel my heart beat fast when I encounter something for the first time. The fortune of finding a new species is unlikely to fall to me.

The discovery of this prawn in his youth would later inspire Emperor Showa to become devoted to the full-scale study of biology. For the readers' information, with respect to the association between Numazu and the prawns, in 1930, Kiichi Nakazawa, a crustaceanist, visited the Numazu Villa to deliver academic lectures on deep-sea organisms to Emperor Showa; he made a significant contribution to the academic and fishery-industry communities through the establishment of the "Suruga-Bay Marine Biological Laboratory" at Kanbara, Shizuoka, at his own expense, and also through his researches into deep-sea organisms, especially a commercially important deep-sea shrimp, *Sergia lucens*.

A short time before the Emperor found the *Sympasiphaea imperials*, a precious message arrived conveying that Princess Nagako, who was the eldest daughter of Prince Kuni Kuniyoshi, had been informally appointed as the Crown Princess. Empress Teimei liked her, and the results of her physical examination showed that she was in excellent physical condition. However, the records of physical examination kept in Gakushuin revealed that the Prince Kuni family had inherited the color-blind gene from the Shimazu family (a daimyo in Kagoshima). In 1920, upon hearing of this fact, Aritomo Yamagata, a Genro in the Meiji period, cultivated a movement in vehement opposition to this appointment and maneuvered to force the Prince Kuni family to decline the nomination for the Crown Princess. This became a so-called "Certain Serious Incident in the Imperial Court." The reason for

Yamagata's opposition was fear that the inheritance of the color-blind gene by the Imperial grandchild might lead to a serious situation. The Imperial grandchild would be destined to command the Imperial Japanese Army and Navy as His Majesty the Generalissimo when he, the Crown Prince, succeeded to the throne in the future. Any man suffering from color-blindness would not pass the physical examination for conscription. This certainly seemed to be the reason behind the rivalry that sprang up between Yamagata, a senior statesman of the Choshu Domain (the current Yamaguchi), and Shimazu, the feudal lord of the Satsuma Domain (the current Kagoshima). For the readers' information, we learn in biology class that color-blindness is characterized by sex-linked inheritance; specifically, if a female inherits this type of gene, color-blindness may develop in half of the male infants born to her. It goes without saying that not all of the females born into such a family line will inherit this type of gene. With respect to this incident, at first, Yamagata's side seemed to be winning, but the efforts made by Jugo Sugiura and the Prince Kuni family ultimately bore fruit, leading to the failure of Yamagata's scheme. The next year, every newspaper reported that no change had been made to the informal engagement. At the beginning of 1921, Yamagata passed away, and soon, an Imperial sanction was given (Kaoru Ono, "Certain Serious Incident in the Imperial Court" [51]; Yukio Ito, "The Life of Emperor Showa" [29], and others). According to informed sources, the fact that Crown Prince Hirohito, who was a biologist, ultimately said, "I agree with the Imperial sanction appointing Princess Nagako as the Crown Princess," was a decisive factor in the final appointment.

Crown Prince Hirohito as Sessho (Regent to Emperor Taisho) and the Establishment of the Biologogical Laboratory of the Imperial Household

In 1921, the Crown Prince, after completing a series of education curricula at the 'Study,' visited European countries for half a year from the beginning of March to the beginning of September as his graduation trip, because the Genros thought that the future Head of State should not be sheltered from the world, but rather unquestionably needed to visit various foreign countries, to observe their cultures so as to enrich his knowledge, and to meet the Heads of State and dignitaries of the various governments for the purpose of building better relationships with them. This is likely something that the Genros, who had experienced the upheaval from the Edo Period to the Meiji Restoration, felt very strongly about. In fact, Emperor Taisho had hoped very much to visit the European countries in his Crown Prince days, but his wish had not come true, owing to various factors. Visits to European countries, as well as to the US, had been considered when the agenda for his foreign travels was being mapped out, but they were not realized at that time. Initially, he visited only two European countries, the United Kingdom of Great Britain and Ireland (the current United Kingdom of Great Britain and Northern Ireland, hereafter simply referred to as the UK) and France. Later, he was to visit three more countries, Belgium, the Netherlands and Italy. The Crown Prince made his trips to and from these countries by way of the battleship Katori, a ship of the Imperial Japanese Navy, accompanied by the battleship Kashima, which was the same type as the Katori. It is said that the attendants on the ship engaged in continuous efforts so that the young Crown Prince, who had never before left his homeland, might acquire European-style manners and etiquette. He also studiously practiced the speech he was to deliver. Thanks to his dedicated effort, his words and behavior were received favorably in the countries that he visited.

With an Anglo-Japanese Alliance being in effect at that time, it was natural that he would spend the lion's share of his time away in the UK, and, accordingly, he seems to have acquired a lot of knowledge there. During a visit to Buckingham Palace, the Crown Prince had the valuable experience of suddenly finding himself in the presence of King George V, and the two had a friendly hour-long talk. Emperor Showa spoke of his memory of that experience at a press conference held at the Nasu Imperial Villa decades after his visit; he talked with George V of the so-called sound way of maintaining a functioning constitutional monarchy in the UK and, since then, he has always kept the discussion in mind. He also visited the British Museum (the present Natural History Museum, London), considered a temple of natural history, and both Cambridge and Oxford Universities, each a proud bastion of time-honored tradition. The University of Cambridge and the University of Edinburgh each awarded him an honorary doctoral degree in law. Moreover, he visited the Linnaean Society in London. This society, established in 1788, is the oldest biological society in the UK, and was where Charles Darwin made his presentation on evolution.

In France, in addition to Paris, on the suggestion of George V, he visited such battle sites as Verdun and Somme, where the scars of World War I were still evident. This experience served as a trigger for the Prince to develop the strong belief that one should "never get involved in any wars at all." Furthermore, he visited Brussels, Amsterdam, Naples and Rome to meet the members of the individual royal families and the Pope, as well as others. Through the thoughtful arrangement of officers of the Italian Government, who were familiar with his preference for animals and plants, while in Naples, he got the chance to visit the Blue Grotto situated on the Isle of Capri, observing the fish and shellfish living in the highly-transparent shallow sea (Masaru Hatano, "Report of the Crown Prince Hirohito's Visit to Europian Countries" [17]).

Also in Naples are the world-famous Naples Zoological Station and its adjunct aquarium. This station was founded by Anton Dohrn, a German zoologist, in 1873, and was operated by means of a unique system in which its running cost was covered by the admission fees charged for entering the aquarium, the world's largest water tank at that time, as well as the rental fees for research tables assigned to the individual countries (at present, it is covered by the financial assistance of the Italian Government). Japan has held one of the research tables in this station since the prewar days and large numbers of Japanese scientists, including myself, have visited the laboratory to study marine organisms. Many Nobel Prize winners have also visited there, including James D. Watson, who is well known for his discovery of the DNA double helix. A survey conducted several years back, by Yukio Yokota, a professor emeritus at Aichi Prefectural University, in cooperation with Giorgio Bernardi, the president of the station at that time, demonstrated that Crown Prince Hirohito had visited this station and that the local press had released an article reporting his visit. "Kyu-chu Monzen Gakuha" ("Observers of the Imperial Court 'School'"), edited by Sukemasa Irie, also indicates as much [24], which would seem to prove this fact, although, in "The Annals of Emperor Showa," it is stated that the aquarium was closed on that day [21]. His visits to the British Museum and the Naples Zoological Station must had given a great incentive to the Crown Prince, given his interest in insects and shellfish, as well as the visits paid to the Misaki Marine Biological Station since his childhood days, to start his research on marine organisms.

Aside from this, Hachiro Saionji, Kinmochi's son-in-law, is known to have accompanied Kinmochi when he visited France to attend the Paris Peace Conference, which resulted in the Treaty of Versailles, written after the end of World War I, as an ambassador extraordinary and plenipotentiary. Hachiro served the Crown Prince as a general affairs official of the Board of the Crown Prince's Household. He, a master of Judo and Kendo, was rumored to be a spirited man and to have gotten into a terrific fight with several ruffians who invaded his home, shouting objections about the Crown Prince's visit to the European countries, and defeated them, despite being injured. Hachiro was the person who taught the 16-year-old Crown Prince and his younger brothers the basics of playing golf. He attended the Crown Prince as a master of ceremonies when the Crown Prince visited Europe. On the battleship Katori, to relieve the Prince's feeling of tedium, a sumo tournament (although another source claims it was a judo tournament) was held per the Crown Prince's preference for engaging in the sport. As the story goes, while most of the competitors intentionally threw their matches against the Prince, feeling it was their duty to do so, Hachiro threw the Crown Prince onto the dohyo (sumo ring) several times with no constraint. It is said that he also eschewed discretion in playing card games, such as bridge and poker, with the Crown Prince. Hachiro would later participate actively in collecting the marine organisms at the Hayama Villa, but passed away in the year following the end of World War II. He was the eighth son of Motonori Mohri, the last daimyo of Choshu Domain, and Goro Mohri, my grandfather, was one of Hachiro's elder brothers.

The Crown Prince, after returning from Europe, was responsible for official duties on behalf of Emperor Taisho, who had been ill for much of his five years as regent. Meanwhile, on September 1, 1923, the Great Kanto earthquake, the most disastrous earthquake in Japanese history, struck the Tokyo metropolitan area directly, causing much serious damage and resulting in more than a hundred thousand deaths and missing persons in total. The Regent made a direct inspection on horseback of the miserable state of the affected area on September 15, and, the following day, informed Nobuaki Makino, the Minister of the Imperial Household Department at that time, of the postponement of his wedding ceremony, scheduled to be held within the year. Makino was deeply moved by the regent's consideration. Thus, the wedding ceremony was postponed to the next year.

In the wake of his return from Europe, and prior to his Imperial marriage, the Regent devoted himself to playing sports such as golf and tennis, as well as horseback riding, all of which were growing in popularity in Europe, with Crown Princess Nagako. In fact, in 1922, the year after his visit to the UK, the Regent played a golf match with Prince Edward, the Prince of Wales (Edward VIII, who later renounced the throne due to his desire to marry Wallis Simpson and became the Duke of Windsor), who was visiting Japan to return the courtesy, on the golf course constructed in Shinjuku gyoen (an Imperial garden) for the exclusive use of the Imperial Family. At that time, Empress Teimei, the Regent's mother, expressed concern about his lifestyle, indicating that she felt that the Crown Prince was devoting too much time to playing sports (Yukio Ito, "A Life of Emperor Showa" [29]). In those days, lectures on various subjects, such as the constitution and morality, were also regularly delivered to the Crown Prince; however, it took a little more time before he began to commit to studying biology on a daily basis.

While at ease in serving as the Regent to the Emperor, the Crown Prince did not begin to study biology in earnest until the end of the Taisho Period. Thus, in 1925, a simple wooden one-story "Biological Laboratory of the Imperial Household," with an area of about 148.5 m^2, was built in the garden of the Temporary Crown Prince's Palace (the former Akasaka Rikyu and the current State Guest House) where he lived at that time, and Hirotaro Hattoti was invited to serve as an official to the Crown Prince and appointed as the director of the "Biological Laboratory of the Imperial Household." He taught the Crown Prince how to use a microscope and how to make prepared specimens of cells and tissues as a first step. Empress Teimei gave him, as a wedding present, a microscope set with accessories made by Leitz in Germany. A large number of specimens prepared for study at that time are now displayed in "the Showa Memorial Institute" (Tsukuba, Ibaraki), the National Museum of Nature and Science. The sections of the specimens are neatly arranged on glass slides, reflecting the scrupulous personality of Emperor Showa, who did everything carefully, step by step. My classmates and I also fashioned this type of prepared specimen in a lab course at the university, but our skills were far inferior to the Crown Prince's. Be that as it may, the research that the Crown Prince engaged in here was no better than that which he did on general biology. In addition to the experimental room, this "Laboratory" had a library and instrument room, a preparation room, and a culture room, as well as breeding and experimental farms. The Laboratory was later destroyed by the fire during the Great Tokyo Air Raid and only its foundation remained unburned.

The Regent, owing to his official duties, engaged in his research here for only limited hours on Saturday mornings or in the afternoon. In the spring following the establishment of the "Biological Laboratory of the Imperial Household," photographs were published of the Regent at work on his research (Fig. 1.5). At this time, it was no secret that the Crown Prince was doing research on biology. After that, the Regent began to make researches on hydrozoans, the subject of his life's work as Emperor Showa, and on myxomycetes, another subject in which he was very interested. According to the book written by Prince Hitachi, a list of specimens shows that the Regent collected myxomycetes for the first time in 1926 (Prince Hitachi Masahito

Fig. 1.5 Emperor Showa at the Biological Laboratory of the Imperial Household in the Akasaka Palace around the end of the Taisho Period (Courtesy of the Imperial Household Agency)

[55]). That year, the Nasu Villa was built for the Prince Regent and the Princess Regent, as described later. Emperor Showa writes that he began his research on hydrozoans in 1929. However, chamberlain Osanaga Kanroji (the later deputy grand chamberlain and chief ritualist of the Imperial Court) describes in his book that the research into both hydrozoans and myxomycetes began in 1926 [33]. It is believed that he began his research on other marine organisms at around the same time.

In December 1926, Emperor Taisho passed away in a house attached to the Hayama Villa, and the Regent, who had stayed in the Villa to nurse the Emperor, was immediately enthroned. The name of the era, Taisho, was thus changed to Showa. This happened when Emperor Showa was 25 years old.

Why Did Emperor Showa Select Biology as His Research Field?

As described earlier, the Crown Prince was very interested in both history and biology when he was studying in the Crown Prince's Imperial 'Study,' ultimately choosing to dedicate his private time to the research of biology. It is said that elder statesmen, such as Togo and Saionji, made efforts to steer the Crown Prince away from too much involvement with historical science, fearing that it might merely cause problems for the future emperor. (The officials of the Imperial Household Agency also made similar considerations when the present Emperor Akihito was deciding on his course of study at university.) However, Torahiko Nagazumi, one of his fellow students, later noted that this decision was possibly made at his own discretion, saying, "I think that the Crown Prince probably refrained from studying history for reasons other than it being undesirable" [47]. Nevertheless, it may be inferred that the officials who served the Prince had some influence on his decision.

To be sure, some may speculate that certain politicians and soldiers took advantage of the Crown Prince's preference for biology to avert his eyes from actual political and military affairs. However, Munetetsu Tei, the author of "Why Do Emperors Make Research on Biology?" [61] recently asserted that the attendants who served the Crown Prince, including Kinmochi Saionji, encouraged him to study biology as part of Imperial governance from the viewpoint of a "national strategy." His position may be summarized as follows:

Among royal and noble families in Europe at that time, it was a mainstream tendency to enjoy the study of natural history (biology) as either a hobby or as a genuine field of study. Saionji, who had studied in France for an extended period, must have had intimate knowledge of this foreign practice. Against this background, detailed knowledge of biology was essential for the purpose of making successful Imperial diplomacy (intercommunication) with the royal families of the European countries, including the UK. This might just be the reason why Saionji encouraged the Crown Prince to study biology. Also, the author of the above-mentioned book guesses that Prince Akishino might have changed his major from political science, which he had studied up to that point, to biology while he was staying at the Graduate School, University of Oxford, for the same reason as that described above.

As described at the beginning of this book, in fact, the royal and noble families of Europe tended to place high value on rare animals and plants, together with various treasures collected from their own colonies, and they even held events such as the "Evening for Microscopic Watching," in which the participants had a chance to observe a wide variety of organisms invisible to the naked eye. It goes without saying that such events might attract tremendous interest among these families; however, it is difficult to believe that biology alone was positioned at the center of academic topics considered indispensable for Imperial intercommunication. Yuzaburo Kuratomi, the President of the Privy Council, cites, in his own diary, Saionji's statement on his involvement in this matter, touching on the implications of the underlined words from various viewpoints:

> At present, the Crown Prince makes research only on the morphology of fish and other animals, and has not yet proceeded to philosophical theory (underlined by the present author).

Stating his opinion that court rituals and biological research were essentially integrated with each other, Takeshi Hara also indicates that this sentence had an implicit influence on the Prince's attitude about court rituals (Takeshi Hara, "Emperor Showa" [16]). However, since the Crown Prince had just begun his research in 1928 when this description was made, it may be considered that this sentence simply meant his study had not yet reached the academic level. For the readers' information, the doctorate of philosophy (Ph.D.) is generally conferred upon those who finished their doctoral courses at European universities. It was later indicated that Prince Akishino was also already in preparation for a change of his major to biology in his student days at Gakushuin. This view is worthy of attention, however, I believe it is not a subject that we need to go any deeper into.

Now, I will introduce my own modest experience in association with this topic. I have always been interested in various forms of life and, as a boy, I became familiar with insects. However, since my father was a military officer, I entered Army Cadet School to be trained as an army officer during World War II, which represented quite a turnaround from my daily life of pursuing insects. Meanwhile, after the defeat, I returned to the middle school where I had been studying previously and then entered high school under the old education system. With this transfer as a start, I was driven by the desire to go forward along the path that I found most favorable, and majored in zoology at the university, despite the advice of Ichiro Tomiyama, a teacher from my high school, who advised me to "give up proceeding along such a path; otherwise you will not be able to make a living." These are the processes that led me along this path up to the present day. Many students who end up majoring in biology have similar stories, especially those who major in taxonomy [43].

On the other hand, with the recent advent of molecular biology, it is not only those who love living organisms who are advancing their academic researches in this field. Physicists and informatics scientists have also entered this field, one after another, suggesting that, to understand biological phenomena, cooperation among those with a wide variety of backgrounds is essential. As will be described later, in such traditional areas of natural science as taxonomy and phylogeny not only have morphological features and ecological varieties frequently been adopted, but molecular biology, in regards to sequences of four types of base, which compose the DNA that carries genetic information, is also actively used for clues. In fact, Emperor Akihito and Prince Akishino have also advanced this kind of research, and the 29th International Prize for Biology (2013) was awarded to Joseph Felsenstein, who had developed new software for estimating a phylogenetic tree of all organisms.

Now, let's get back on track. Nagazumi, who was one of Emperor Showa's fellow students and had served the Emperor for a long period, describes the reason why Emperor Showa selected biology as his research subject as follows:

> I think Emperor Showa selected biology at his discretion, and that no statesman was involved in his decision. However, it is considered that, whereas the advice provided by Prof. Hattori might have had some influence on his selection of taxonomy among the biological sciences, he was meant to walk the path of biology because he was inherently interested in shellfish and insects [47].

It is logical that the memories and views of his fellow students and the chamberlains who served Emperor Showa, who were closely involved in the Emperor's daily life, would reflect the true nature of Emperor Showa far more exactly than those in the outside world. At a press conference held prior to the Commemoration Ceremony of the 50th Anniversary of His Majesty the Emperor's Accession to the Throne, Emperor Showa also stated:

> I have been interested in the collection of shellfish and insects since my childhood days and enjoy it as a hobby now, just as in the saying, "What is learned in the cradle is carried to the grave." I was interested in affairs related to war history and political history during the period of learning at the Crown Prince's Imperial 'Study.' My knowledge of these affairs might run the risk of being exploited by persons with different ways of thinking, and, to engage in research on historical science, I might have to sit at my desk for a long time, just as in Zazen practice, potentially harming my health. Thus, I considered biology to be appropriate for my circumstances and decided to walk along the path of making detailed research into it.

Emperor Showa simply followed an orthodox method, by which he watched and collected animals and plants in the fields, and this contributed to his better health and stable emotional state. It may be possible to speculate about this matter, but it seems to be correct that he selected his favorite subject at his own discretion, enjoyed it as a hobby and, finally, enhanced his ability to further his research up to the academic level.

Emperor Showa Decides to Devote Himself to the Study of Taxonomy

The coronation of Emperor Showa, who had acceded to the throne, was held in Kyoto in 1928. This was just after Zuolin Zhang, who was the leader of the Munchurian Military Clique, had been killed with a bomb by the Kwantung Army in Manchuria (northeast China), and right in the middle of the Great Depression. Against this background, Japan itself, along with the young Emperor, were put into a difficult situation. After that, Japan was confronted with unparalleled difficulties, one after another: the Manchurian Incident, the May 15 Incident, the February 26 Incident, the Second Sino-Japanese War and the Pacific War (World War II).

As of 1930, the daily workload of Emperor Showa was summarized as follows:

Monday: Received lectures delivered by experts in individual specialized fields;

Tuesday: Received lectures on public administrative and fiscal affairs;

Wednesday: Received an audience of diplomats and attended to his official duties;

Thursday: Received lectures on the Imperial Household Law system and military affairs;

Friday: Received an audience of those admitted, including diplomats; and,

Saturday: Received lectures on biology from Hirotaro Hattori

(Ito [29]).

Among his tours after World War II, Emperor Showa regretfully had to cancel a scheduled visit to Okinawa due to the development of pancreatic cancer late in life. However, the Emperor visited all corners of the country, including Karafuto (present-day Sakhalin) and Taiwan, which were Japanese territories at that time, in addition to Okinawa for site inspection in his Crown Prince days. In 1927, he visited the Ogasawara (Bonin) Islands for supervision of a large navy training exercise on the battleship "Yamashiro". For first two days, he stayed on Chichijima Island and, on the next day, landed on Hahajima Island with Hirotaro Hattori so as to collect marine organisms. Some hydrozoans collected at Hahajima Island at that time were later detailed in an article entitled "Some hydrozoans of the Bonin Islands." This suggests that he had begun part of his study on hydrozoans earlier than 1929, as chamberlain Kanroji states in his book [33].

In Ogasawara, the two chamberlains, Saionji and Kanroji, both of whom were good swimmers, dived into the surrounding sea areas with local fishermen to collect marine organisms. At that time, since the launch that Emperor Showa was on had been already been filled with organisms that they had collected earlier, they began to load the new collection on to an accompanying boat. Subsequently, a young naval officer on that boat barked at them within their hearing, "This boat is accompanying the battleship for reasons other than carrying cargo," as Osanaga Kanroji states in his book [33]. This story seems to reflect the idea that "the Crown Prince, even though he will become an emperor in the future, feels enthusiasm for birds and beasts, as well as plants…, we wonder about his attitude. He should learn military affairs and politics and economics for a better national government." This idea had smoldered in the military, especially in the army, throughout the war. Whereas Hattori served as a guard to protect him from such criticism, saying, "It is worthy of praise that Emperor Showa gives attention to even birds and beasts. I hope that His Majesty will select his favorite subject at his own discretion" (Tei [61]). Kanroji and Saionji also actively encouraged him to devote himself to the study of biology, saying, "It is a preferable hobby." That year, he visited Amami Oshima Island, and two years later, Hachijyojima and Izu Oshima Islands, to collect marine organisms and plants. It goes without saying that I have gleaned only the information on living organisms from a number of different publications concerning Emperor Showa and that his collecting activities took place in the intervals between his official duties.

Hattori, as head of the Biological Laboratory of the Imperial Household, encouraged Emperor Showa's study of myxomycetes, his own subject of study, as well as the study of marine hydrozoans. At least at that time, the general public was unfamiliar with both of these sorts of organism. Each of the organisms will be described a little later. Answering the question, "Why did he (Prof. Hattori) encourage Emperor Showa in the study of taxonomy?" Nagazumi stated:

> Did Prof. Hattori say so? Taxonomy is such an inherently academic field that young researchers do not always actively intend to study it, though it is one of the fundamental disciplines. For this reason, taxonomists often have only a few chances to spur debate in the academic world. I guess that Hattori gave modest consideration to such a situation. Taxonomy allowed Emperor Showa to collect and keep the specimens that he found. I speculate that Hattori also considered that, were the Emperor to select taxonomy, he would be able to study the previously collected specimens in the free time between his official duties. It is requisite that he use his off-time for studying [47]. Emperor Showa himself probably felt the same way as Hattori.

Answering a question posed by Toku Tanaka, a journalist from Kyodo News who closely watched the Emperor's research activities on biology throughout the mid-war to post-war period, Emperor Showa stated:

A lot of researchers are engaged in the applied studies, while a few specialize in taxonomy, so I selected the latter. Recently, I have also been doing research on a wide variety of organisms, including hydrozoa, as well as on microorganisms, molluscs, coelenterates and bacteria. I intend to continue this research throughout my life.

Moreover, Tanaka states, in his book entitled "Emperor Showa and his Research on Biology," [60] issued in 1949:

This is a research subject that no one enters into expecting to be beset by rivals, but someone will need to engage in it in the future.... This was probably the position taken by His Majesty and reflects his consideration.

Later, at the press conference held prior to the Commemoration Ceremony of the 60th Anniversary of His Majesty the Emperor's Accession to the Throne in 1986, Emperor Showa stated:

I selected research on hydrozoans on Hirotaro Hattori's advice. I thought that, as few researchers were engaged in taxonomy of hydrozoans in Japan at that time, I might be free from competition with other researchers.

The reason why Emperor Showa chose to walk the path of the taxonomy of myxomycetes and hydrozoans seems obvious from the above words.

Hatsuki Tsujimura, a zoologist who had worked at the Biological Laboratory of the Imperial Household and served Emperor Showa as a chamberlain for a long period from the time shortly after the end of World War II, had his own speculation, as described below. Emperor Showa probably selected hydrozoans as the focus of his zoological interest because (1) the polyps (see the next section) of hydrozoans are most suitable in size for preparing specimens with no need for a large-scale research facility, making it possible to rear them, and (2) there are a large number of species whose life history, including correspondence between polyps and jellyfish, has not yet been revealed (Naohide Isono, "Those Who Came to and Went from the Misaki Marine Biological Station" [27]).

What Is a Hydrozoan?

Now, let's turn our eyes to the biology of hydrozoans. Hydrozoans, including Jellyfish and sea anemones, belong to the phylum Cnidaria (formerly called the phylum Coelenterata). The name Cnidaria (cnidarians) is derived from the stinging cells (called cnidoblasts or nematocysts) on their tentacles, used for capturing prey and defending themselves against predators.

As is known from images of them, which can often be seen on TV, comb jellies are so similar to jellyfish that we find it hard to distinguish between them. However, they are entirely different from the jellyfish of the phylum Cnidaria, in that they have eight lines of giant cilia, called comb-plates, which are used for swimming, and, accordingly, they are classified by another phylum, Ctenophora.

To digress momentarily from the subject, since Carl von Linné, the renowned Swedish biologist, established the foundation of modern taxonomy, organisms have largely been classified downward into kingdom, phylum, class, order, family, genus and species, according to the taxonomic hierarchy. Moreover, they are further sub-classified into intermediate groups, subphylum, suborder, subspecies, and so on. By the way, the taxonomic position of humans (*Homo sapiens*) can be followed downward as shown below:

Kingdom: Animalia, Phylum: Chordata, Class: Mammalia, Order: Primates, Family: Hominidae, Genus: *Homo*, and Species: *Homo sapiens*.

The most familiar species of hydrozoan may be the small fresh-water *Hydra*. This animal is, in appearance, similar to but simpler than a sea anemone, and has the capability to regenerate lost fragments of its body, even if they have been chopped off. For this reason, species of *Hydra* have often been used in biology and other classes as materials for experiments. This sessile form (body plan) in cnidarians, called a "polyp," has a stick-shaped (stalked) body (generally, cylindrical) with a mouth opening at its tip, surrounded by tentacles. They attach to aquatic plants in fresh water using their pedal disks at their aboral ends. Hydrozoan polyps generally reproduce asexually by budding, but may produce sperm and eggs for sexual reproduction in some cases. The Emperor Showa mainly studied marine hydroids, many of which have the Japanese name "umi-hydra (marine hydra)." Most of them have a plantlike appearance, with a lot of small polyps (hydranth) on their branches, and they form into colonies, as seen in corals, and adhere to substrata such as rocks. All of the organisms described above belong to the class Hydrozoa (Fig. 1.6).

Many species of Hydrozoa alternate between two life forms in their lifecycle, polyp (asexual generation) and medusa (jellyfish, sexual generation). Many live as either polyps or medusae throughout their life. Some produce medusae, which, after detaching, regenerate sexually through fertilization. Fertilized eggs grow up into planula larvae and, after separating from their parents, fix onto the substrata as polyps. Medusae and polyps seem to belong to different species; however, were one to try to pick up a polyp and then turn it upside down to open its umbrella, it would assume the form of a medusa. Otohime-no-Hanagasa (*Branchiocerianthus imperator*), a large solitary hydrozoan species about one meter in full length, was collected off of Misaki for the first time. Later, the Emperor Showa compiled the specimens of two suborders of the order Hydroida (Athecata and Thecata) into monographs. The class Hydrozoa also include other orders, such as hydrocorals, which form a calcareous colony, and Siphonophorae, including the blue *Physalia physalis* (Portuguese Man O' War), with its quite venomous nematocysts that cause a sharp pain in those who are stung by its tentacle, leading to death in the worst cases.

Athecata hydroids are characterized by naked hydranths with no perisarc and contain many species with no medusoid stage, such as the *Hydra* and *Hydractinia* species, including *Hydractinia epiconcha*. Some species of Athecata, such as

Life cycle of a hydrozoan *Obelia*

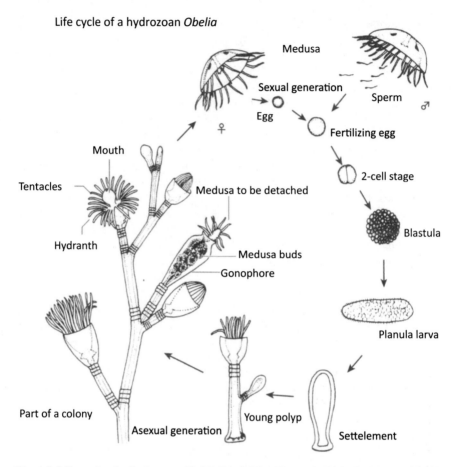

Fig. 1.6 Life cycle of a hydrozoan *Obelia* (From "The Emperor's Biological Research" [63], courtesy of the National Museum of Nature and Science)

Spirocodon saltatorix, however, produce beautiful medusae. I often found examples of this animal in Aburatsubo Bay in Misaki, Miura Peninsula, in my college days. The Thecata species is characterized by hydranths enveloped within the perisarc and medusa growing within the gonotheca. Plumulariid species of Thecata form large solid colonies, which resemble plants in appearance and have Japanese names derived from plants, such as xxxx-"kaya" (torreya) or xxxx-"hiba" (cypress). A lot of species belonging to other Thecata families form small colonies and produce medusae.

    We may notice that, among many species, the hydrozoans are fantastic organisms if we take a closer look at them, even if hydrozoans may be less noticeable than shellfish, naked sea slugs (Nudibranchs), and shrimps and crabs, the biological investigation and classification of which the Emperor Showa entrusted to other researchers.

In Japan, Masaharu Inaba was the first researcher to turn his attention to hydrozoans as the subject of his study (Mayumi Yamada, "Biologists, who were devoted to researches on hydrozoans" [70]). He graduated from the Department of Zoology, Faculty of Science, University of Tokyo, in 1889. On the suggestion of Kakichi Mitsukuri, the first Japanese professor of the Department of Zoology, he made observations of a wide variety of marine organisms at the newly-constructed Misaki Marine Biological Station. Among these organisms, he was so specifically fascinated by hydrozoans that he contributed an article on the specimens that he collected in Misaki, Shima (the current Mie) and Kii (the current Wakayama) to "Dobutsugaku Zasshi" (Zoological Magazine, later renamed "Zoological Science") soon after it was first published. At the beginning of his article, he wrote, "In the past, almost no research on the order Hydroida has been undertaken in Japan...," based on which it was guessed that he might engage himself in research on hydrozoans through reference to foreign publications, literatures, papers, and so on, so that he might publish his articles. Inaba, who was born into a temple family, quit as a researcher after graduating from the University, and later served as the president of Otani University for Buddists in Kyoto.

In 1904, Franz Doflein, a German zoologist, visited Japan and conducted a comprehensive survey on the marine organisms in various parts of Japan, especially in Sagami Bay, which lies south of Kanagawa Prefecture. During this survey, the zoologist who was responsible for the detailed investigation of hydrozoans out of the collected biological specimens was Eberhard Stechow. At that time, he worked at the Bavarian State Collection of Zoology in Munich ("Fauna Sagamiana" by the National Museum of Nature and Science [62]). Born in 1883, he was a disciple of Richard von Hertwig, a renowned German zoologist. Stechow put the findings of his research on the hydrozoans collected in Japan into a paper, with reference to the article written by Inaba, which was later translated into English by Seitaro Goto, who was one year behind Inaba at the University of Tokyo and later worked at the University as a zoology professor. Stechow also collected specimens of hydrozoans from various parts of the world and compiled an Identification Key that would serve as the basis for classifying the collected specimens into any of the families and genera. The Emperor Showa wrote, in his first article on the hydroids of the family Clathrozonidae, of which he made a study, that he had failed to classify the hydroids according to Stechow's classification system.

Until the breakout of World War II, the Emperor Showa had been in the habit of sending the collected specimens to biologists such as Elof Jäderholm, Sweden, Charles McLean Fraser, Canada, and Eugène Henri Joseph Leloup, Belgium, in addition to Stechow, in order to get answers to his taxonomical questions. Looking back on those days, Mayumi Yamada, a professor emeritus at Hokkaido University, who edited the unpublished monograph written by the late Emperor Showa, told of how the reprints of papers written by Stechow had been severely damaged due to extensive use and repeated rebinding by the Emperor. After the end of World War II, the Emperor had the opportunity to talk with several foreign biologists who specialized in hydrozoans, but not with Inaba or Stechow at all.

As known from the above-mentioned context, at the time that Emperor Showa began his research of hydroids, there was no biologist in Japan who specialized in the class Hydrozoa in the true sense of the word (Yamada [70]). Also, in the preface of the monograph "The Hydroids of Sagami Bay," the Emperor stated that:

> I initially began my study of Hydrozoa on the advice of the late Dr. Hirotaro Hattori. During the first twenty years, I continued my research under his guidance; as of 1945, I continued with the advice of the late Dr. Ichiro Tomiyama, as well as the assistance of Mr. Hatsuki Tsujimura. During this period, the late Drs. Tadao Sato and Kenzo Kikuchi also offered their valuable counsel.

Kenzo Kikuchi, an ecologist, was the son of Dairoku Kikuchi, who served successively as the Minister of Education and the President of the University of Tokyo. After graduating from the Department of Zoology, University of Tokyo, he worked at the Otsu Hydrobiological Station, Faculty of Science, Kyoto University (the former Kyoto Imperial University). During World War II, he served as an assistant professor (the official title was recently changed to "associate professor") at the Misaki Marine Biological Station. He was appointed as a professor after the end of the war, and passed away soon afterward. He also served as goyogakari (a general affairs official), a position that was taken over by his successor Tomiyama.

Additionally, the names listed in the preface of the paper written by Emperor Showa include Toru Uchida, a professor of Systematic Zoology, Hokkaido University, who was a researcher on cnidarians (coelenterates), especially jellyfish, as an advisor, and Mayumi Yamada, a successor of Uchida, as a reviewer of the findings of the Emperor's research. As an aside, it is to be deplored that, today, only a few biologists engage in research on hydrozoans in Japan, including Hiroshi Namikawa, who manages the specimens collected by Emperor Showa at the Showa Memorial Institute in the Tsukuba Branch of the National Museum of Nature and Science.

Study of Myxomycetes and the Fruitful Encounter with Kumagusu Minakata

Now, let us move on to another topic, myxomycetes. Myxomycetes are referred to as true slime molds, or simply as slime molds. These shapeless masses of vivid colors such as red and yellow often leave a mysterious impression on us when we find them on dead leaves and decaying logs. These are the just plasmodia—that is, bodies of the growth stage of myxomycetes. The plasmodium, a multinuclear protoplasm, exhibits ameboid movement and feeds mainly on bacteria, decomposing plant remains and fungi. Once it grows up, it extends a branched network of veins, inside which rapid protoplasmic streaming occurs. At maturity, the plasmodium separates itself into a number of different masses, each of which changes into a very small, mushroom-like fruit body. The nuclei within the fruit body are divided by meiosis to form haploid spores. From each of the germinating spores, a

Fig. 1.7 *Physarella oblonga*, one of the slime molds (myxomycetes) discovered by Emperor Showa; its plasmodium and fruit bodies are shown in the upper and lower photos, respectively (Courtesy of Masana Izawa)

mononuclear myxoamoeba or biflagellate zoospore is discharged and, in turn, mates with a male or female one to form a diploid amoeba. The diploid amoeba repeats nuclear divisions to grow into a larger multinucleate plasmodium, thus achieving its life cycle (Fig. 1.7). *Physarum polycepharum*, one myxomycete species, lives in dark and moist places, such as piles of fallen leaves and decaying logs on a forest floor, and grows up into a large yellow multinuclear plasmodium. This myx-omycete is commonly used in a wide range of researches as a model organism, because its plasmodium is easy for researchers to cultivate in a laboratory environment.

As described earlier, Emperor Showa collected myxomycetes in the Nasu Imperial Villa, newly-built in 1926. In the same year, Kumagusu Minakata, a prominent researcher on myxomycetes, offered ninety specimens of myxomycete, which he himself selected for the Emperor through his disciple, Shiro Koaze. The

Emperor collected myxomycetes for the first time at the Akasaka Palace in 1925, and then in various places such as the Imperial Palace, the Nasu Imperial Villa, the Nikko Tamozawa Imperial Villa, the Izu-Amagi cordillera, and the Jimmuji area near the Hayama Imperial Villa, where myxomycetes were rich in variety. In those days, it was not disclosed to the public that the Emperor was engaging in research on myxomycetes, even though he was publicly known to have been appointed an honorary member of the Linnean Society of London in 1932 for his achievement in research on myxomycetes. Against this background, the articles on his works were presented to different journals by Hirotaro Hattori, Yoshikazu Emoto, a professor at Gakushuin University, and Gulielma Lister, an authority on myxomycetes from the British Museum. The illustrated reference book on his work, "Myxomycetes of the Nasu District," which included 125 species of myxomycete, was edited and published privately by Hattori in 1935. All of these works are very precious sources of information on the myxomycetes of Japan.

In 1929, Emperor Showa visited the Kansai region for a site inspection of the industrial conditions on an Imperial cruiser, the "Nachi." Through the mediation of Hattori, the Emperor took the opportunity to meet Kumagusu Minakata on the island of Kashima inside the Bay of Tanabe, Wakayama, and the two collected myxomycetes together. Kashima's natural environment had been conserved by Minakata. Nevertheless, weeds along the path were yanked in advance by regional officers. According to the weekly "Shukan Asahi," issued on February 7th, 2014, Emperor Showa noticed that the appearance of the land had been artificially sanitized, and said

> I have heard that this island is covered by a natural forest unaffected by human beings, but what I am seeing here is not natural.

Minakata was very ashamed to hear his words. This event gave him the opportunity to encourage further awareness of the conservation of Kashima's natural environment, and he was subsequently successful in getting Kashima designated as a "Special Natural Monument" in 1935.

On the Imperial cruiser, Minakata gave a lecture on myxomycetes and other things in the presence of the Emperor, going 35 min over the scheduled time. It was a well-known story that Minakata presented 110 specimens of myxomycete to the Emperor wrapped in caramel packages. These packages are now on exhibition in the Emperor Showa Memorial Museum, Showa Kinen Park. As described below, Minakata was also known for his oddball character and was called an eccentric or a strange person by many. However, he was granted an audience with the Emperor and was not blamed for using the caramel packages in the presentation of the specimens, though we, the so-called war-generation people, deemed his behavior to have been extremely disloyal and irreverent according to the common sense of that time. This may be because Japan was dominated by a broad-minded spirit that valued scholarship before World War II (Hara [16]). Minakata presented the myxomycete specimens to the Emperor through the intermediary of his disciples on four separate occasions in 1932.

A naturalist and folklorist, Kumagusu Minakata (1869–1941, Fig. 1.8), a man difficult to classify into any of the existing academic areas, entered the Tokyo Yobimon (Preparatory School of the University of Tokyo) with Shiki Masaoka, a famous haiku poet, Saneyuki Akiyama, an admiral, Kotaro Honda, a physics professor, and others. However, two years after admission, he dropped out of the school and went to the United States (hereafter, simply referred to as the US), where he devoted himself to investigations of flora and fauna and to reading a large number of books, successfully discovering a new species of lichen and publishing a paper on lichens in "Nature," an English scientific journal. After moving to the UK in 1892, he worked at the British Museum as an assistant editor of a catalogue of the Oriental books kept in the Museum. Later, he was forbidden from entering the Museum owing to an unspecified incident of violence, and he returned to Japan in 1900. While living in the UK, he had a relationship with Sun Wén, who had fled to London at that time. By the way, Minakata submitted as many as 50 papers to "Nature." This journal is highly esteemed as the most authoritative voice in the field of natural science, on par with the American scientific journal, "Science." Accordingly, it goes without saying that the papers that appeared in "Nature" have been referred to many times, and we researchers could only be so proud if our papers were to be accepted by this journal for publication.

Fig. 1.8 Kumagusu Minakata with his wife (Courtesy of the Minakata Kumagusu Kinenkan)

After returning to Japan, Minakata lived in Tanabe and collected a wide variety of plants and fungi in the surrounding areas, including Kashima, preparing them into dry specimens. Myxomycetes were included in his collection, and he registered as many as 178 additional new species in the list that already contained the existing 36. *Minakatella longifila*, a myxomycete species that he had discovered on a living persimmon tree, was determined to be a member of the genus *Minakatella*, newly established and crowned with his name by Lister. For this reason, there is no room for doubt that he was also an outstanding expert on myxomycetes. Nevertheless, he wrote few papers on his research in this field and daringly chose to serve as a researcher out of office. He had a close relationship with the folklorist, Kunio Yanagida.

Seeing Kashima when he stopped over in Shirahama on the way to Ise in 1962, Emperor Showa remembered that he had collected myxomycetes with Minakata, and composed a waka poem:

> Having a view of Kashima seen dimly in the rain, I feel nostalgic about Kumagusu Minakata, born in Kinokuni.

Kinokuni is the current Wakayama Prefecture. A monument inscribed with the above waka poem stands at the site of Banshoyama, Shirahama, from which visitors can get a good view of Kashima.

The examination made by Prince Hitachi, together with Hiromitsu Hagiwara, an expert on myxomycetes from the National Museum of Nature and Science, and the former chamberlains, showed that the research on myxomycetes conducted by Emperor Showa was close to being finished in 1935, and then resumed in 1947 after World War II. He collected the latest specimens in about 1969. The Emperor had been interested in higher plants starting in 1948 or 1949, and focused entirely on collecting them in 1953. For this reason, only a small number of myxomycete specimens were collected in the few decades after 1953 (Prince Hitachi [55]). Presumably, the methods of fieldwork are different between myxomycetes and plants. The number of new species, new varieties, and newly-recorded species of myxomycete discovered in Japan by the Emperor is too many to count on ten fingers. His achievement in regard to myxomycete research earned him his 1932 honorary membership in the Linnean Society of London.

The Biological Laboratory of the Imperial Household

Following the move of Emperor Showa and Empress Kojun to their Kyujyo (the present Imperial Palace), a new biological laboratory was built to the south of the Fukiage Omiya Palace in 1928, called the Biological Laboratory of the Imperial Household (Fig. 1.9). The wooden, western-style Biological Laboratory of the Imperial Household, which was one-story and covered 825 m^2, was composed of laboratories, a preparation room, a library room, and a specimen room. After World War II, a two-story concrete specimen room was added. In the hallway, a large

Fig. 1.9 The appearance of the Biological Laboratory of the Imperial Household (Courtesy of the Imperial Household Agency)

single-panel screen made of a round slice from the trunk of a camphor tree (*Cinnamomum camphora*) was installed and, in a hall that lay beyond the main hallway, ornaments, including a deer head with antlers, were hung on the wall for decoration. The current reception room to the immediate left seems to have been used as a library room previously. In the preparation room next to the reception room, liquid-immersed specimens, herbarium (pressed-leaf) specimens, and prepared microscope slides were displayed. The first and second laboratories followed the preparation room. The appearance of the first laboratory has been reproduced, down to the instruments and office furniture that were actually used by the Emperor, in the Emperor Showa Memorial Museum, Showa Kinen Park. The desks and chairs, which have been used since the foundation of the Biological Laboratory of the Imperial Household, are all very simple in construction.

Now, with reference to detailed descriptions of a eulogy for Emperor Showa taken from the writings of Edred J. H. Corner of the University of Cambridge for the Royal Society [9], the book "Showa Ten'no Kinenkan" ("The Emperor Showa Memorial Museum") [58], published by the Showa-Seitoku Memorial Foundation, and other literature, let us try to get an idea of what the Biological Laboratory of the Imperial Household looked like in those days. The first laboratory was a rectangular room with two south-facing windows, a small square desk, on which the Emperor might spend time reading and writing, and, next to that, a desk with an Olympus discussion-type microscope on it. This microscope enabled two researchers to

observe one specimen at the same time, to facilitate discussion. There is a photo of
the scene, showing the Emperor and Tomiyama observing a specimen together.
A set of both stationary and rotary bookshelves, housing hydrozoan-related literature, occupied the space in front of the desk used by the Emperor. Next to the
Emperor's seat was another seat, presumably for the director or other specialists to
sit on while working with the Emperor, and in front of those were more bookshelves, these housing dictionaries of foreign languages and collections of the
research notes detailing the discussions of the Emperor's works that took place in
the lab. Cabinets placed alongside the corridor-side wall on the right-hand side
contained more than 13,200 prepared specimens of hydrozoans. About 10,000
among them were prepared by Hiroo Sanada, a specialist, and the remaining ones
were prepared by Tatsuya Shimizu.

An experimental workbench by the window featured a Leitz microscope and a
stereoscopic microscope, which were initially used by the Emperor Showa. The
drawer of the workbench held a set of dissection kits, and writing implements
regularly used by the Emperor, including pencils, an eraser, and an old ink bottle,
were placed on top. At the remote end, to the right-hand side of the workbench,
bookshelves housing academic journals, academic literature, original drawings used
for publications, and other items stood in a row. It is said that a total of 11
microscopes were installed, including a Zeiss binocular microscope equipped with a
photographic apparatus, a gift from Germany. The positions of the microscopes
were occasionally rearranged. In the second laboratory, right next to the first one, a
large workbench was installed, the top of which was occupied by a number of
specimens. The room next to the second research room, which is now used as an
animal room, was a preparation room, according to the book "Emperor Showa and
His Biological Research," written by Tanaka [60]. And in what was then an animal
room, possibly the current dark room at the end of the passage running from the
entrance, hydrozoans were bred.

The specimen room covered two floors of the main building. The specimens of
hydrozoans and other cnidarians (coelenterates), fish and other organisms, and a
large number of specimens of shells and insects were displayed in the spaces on the
first and second floors, respectively. The afore-mentioned specimen cases housing
plant and insect specimens that the Emperor collected during his elementary school
days (at Gakushuin Primary School) were also placed in the latter space. The first
two-story specimen room, which had been newly built, had, on the first floor,
specimen types, which served as references for classification, a collection con-
taining a wide variety of fish and invertebrates, excluding cnidarians, all collected in
Sagami Bay, and books, while the second floor, in addition to more books, featured
taxidermed reptiles, birds, and mammals, dried specimens of myxomycetes and
seaweeds, herbarium specimens of plants collected on the grounds of the Imperial
Palace and around the Nasu and Suzaki Imperial Villas, and dried specimens of
various varieties of rice plant, which the Emperor himself cultivated. In the second
specimen room, specimens of coral were housed. Amazingly, the total number of
specimens was roughly 34,000 for animals, excluding insects, and about 23,000 for
plants. Most of these specimens were displayed and stored in the Showa Memorial

Institute, the National Museum of Nature and Science, and a portion of them were kept in the museum in Showa Kinen Park. The bird specimens were transferred to Yamashina Institute for Ornithology, Abiko, Chiba. To the south of the Biological Laboratory of the Imperial Household lay cultivated fields, including a paddy field, in which the Emperor cultivated rice, and a mulberry field, in which the Empress picked mulberry leaves to feed to the silkworms reared at the Sericultural Station. At present, kiwi fruits are also planted there. It is said that orchids and other flowers given as tributes were cultivated in a greenhouse attached to the Biological Laboratory of the Imperial Household.

Emperor Showa devoted himself more earnestly to his research at the Biological Laboratory of the Imperial Household. During much of the same period, Japan was compelled to be involved in a conflict with the Republic of China; when looking at internal affairs, we can see that the May 15th Incident and the February 26th Incident occurred around this time. These external and internal states of affairs forced the Emperor to make extremely tough calls all of the time. Perusing "Irie Sukemasa's Diary," written by Irie [26], a grand chamberlain for the Emperor at that time, we read that the Emperor went to the Biological Laboratory of the Imperial Household every Saturday morning or afternoon, and occasionally on Monday, when he could find spare moments away from his official duties to engage in research. According to the Diary, the Emperor would sometimes go back and forth between official duties and the Laboratory in the same day. At any rate, the Diary gives a true account of the fact that the Emperor had only limited time that he could give over to research. The Emperor, who made it a rule to give top priority to official duties and court rituals, always prefaced his forays into research by saying to an attendant, "I am going to the Biological Laboratory of the Imperial Household. Is there anything I should do first?", always adding, "Be sure to inform me if anything should happen," even though he knew it might further limit the time he was allowed for research (Kanroji [33]). And, upon receiving the answer that there was "nothing to do" at that moment, the Emperor would go happily out to the Laboratory. And yet, despite his modest behavior, some military attaches to the Palace criticized him unjustly, saying, "It is outrageous that the Emperor would devote so much time to biological research under a situation of such tense foreign affairs" (Hara [16]).

Once the Sino-Japanese War had broken out in 1937, the Emperor, being a man of integrity, gave adequate consideration to the delicate situations of the people around him, ultimately sacrificing his desire to visit the Laboratory to do research when the criticism of the military attaches to the Palace reached his ears. Consequently, the Emperor got sick, partially due to the incredibly rigorous schedule that his official duties concerning the war forced upon him. Kinmochi Saionji, fearing for the Emperor's poor health, gave secret instructions to Saburo Hyakutake (an admiral), the then-Grand Chamberlain, to advise the Emperor to go to the Hayama Imperial Villa for recuperation, but the Emperor stubbornly refused to listen. Not knowing what to do, Prince Kan'in Kotohito, who was a patriarch of the Imperial Family and the chief of the general staff of the Imperial Japanese Army, and Prince Fushimi Hiroyasu, who was the chief of Naval Operations, also tried to persuade him. The Emperor reluctantly succumbed to their persuasion and went to the Hayama Imperial Villa the next year, bringing along a collection of marine organisms (Ito [29]). Not long afterward, the Emperor went back to his research in the Laboratory. This anecdote lends credence to the notion that the Emperor, who was of sincere personality, was able to relieve daily pressure and stress, and restore his ability to make well-balanced judgements, by engaging in biological research.

Thus, he restarted his research after a while, although the outbreak of World War II in 1941 forced him, against his wishes, to give up the pursuance of his study at the Hayama Imperial Villa. He did, however, go to the Nikko Tamozawa Imperial Villa the next year. It is said that he discontinued his study at the Laboratory after the Battle of Saipan, where almost all of the soldiers in the Saipan Garrison of the Imperial Japanese Army died with honor. Reaching back into the memories of my childhood, it seems to me that the people of Japan probably did not know of the close relationship between the Emperor and biology, owing to the military blackout of news. On September 6th, 1941, the basic national policy of "Preparation for war against the US, the UK and the Netherlands" was proclaimed. In this context, the Emperor recited a waka poem written by Emperor Meiji;

I would like the sea surrounding our country in all directions to be calm, so why does it dare to make waves?

(Interpretation: I sincerely wish that all of the countries of the world would respect each other as neighbors. So why is a conflict about to break out?)

The Emperor bemoaned the situation, telling his people, "Although I intend to follow the spirit of the Emperor Meiji, who longed for peace, by continuing to recite this poem, it must be regretfully admitted that the political situation is far from good" (Kanroji [33]), whereupon he then went to the Laboratory. Hearing about this instance, Hideo Kishida, a senior staff writer at *Asahi Shimbun*, when commenting on "Irie Sukemasa's Diary," noted, "the Emperor seems to find peace of mind only when immersing himself in his research" [26]

After World War II, the Emperor was released from various daily restraints, including military duties, and became free to continue his research on organisms; accordingly, this genuine aspect of the Emperor became widely known to the public. In January 1946, the year following the end of World War II, he resumed his research in the Laboratory. Hattori's preparation for this research required three weeks, in part because of the unavoidable interruptions that had occurred due to the war. The Emperor spent time in the Laboratory on Monday and Thursday afternoons and all day Saturday, unless official duties had been scheduled. In the 1950s, the Emperor seemed to arrange his schedule such that he could make researches on hydrozoans on Thursdays and Saturdays, leaving other subjects for Mondays.

Sukemasa Irie wrote about the Emperor's life at that time in his essay "Horibata Zuihitsu" ("Essays around the Imperial Moat") [23]:

Unless official duties had been scheduled, the Emperor went to the Biological Laboratory of the Imperial Household both in the morning and afternoon on Saturday, and worked on the classification of hydrozoans for six hours in total. Also, he observed the samples on the prepared slides under a microscope, discussed the images of the objects projected onto the projection microscope with Ichiro Tomiyama and referred to the appropriate literature. He always wrapped up his research at about 16:30, so that Tomiyama and the staff working at the Laboratory might leave work on time, although the truth was that he hoped to continue his research there late at night.

Irie also states that, after returning to the Imperial Palace, the Emperor worked on a subordinate task concerning "Flora nasuensis," a part of which he wrote himself, during the time period until dinner.

The first director of the Biological Laboratory of the Imperial Household was Hirotaro Hattori, who, as mentioned above, was one of the Emperor's former teaching staff at the Crown Prince's Imperial 'Study.' Starting from the time before the outbreak of World War II, he had given Emperor Showa academic guidance, and after the war, he maintained his position as director of the Laboratory until 1964. The next year, he passed away at the age of 90. It is said that the Emperor referred to Hattori by his title of "sensei (professor or teacher)" all of his life. Ichiro Tomiyama was appointed as Hattori's successor. Tomiyama (1906–1981) had worked at the Laboratory as a taxonomist since 1949, on the recommendation of Yo (Kaname) Okada, the head of the Department of Zoology, University of Tokyo. Tomiyama also served as the director of the Misaki Marine Biological Station (Fig. 1.10). I heard an interesting story about him, admittedly through hearsay. It

Fig. 1.10 Emperor Showa observing hydrozoan samples, with Ichiro Tomiyama, the director of the Biological Laboratory of the Imperial Household (Courtesy of the Imperial Household Agency)

happened one night, during which Tomiyama was helping the Emperor in his research in the Nasu Imperial Villa. The two men were in separate rooms when Tomiyama heard the Emperor talking to himself, and he wondered what the reason for this was. The truth was that the Emperor was practicing his speech for the opening ceremony of the House of Representatives the next day. Tomiyama also said that, when meeting foreign researchers, the Emperor took the international situations of their countries into consideration, though I cannot furnish any details of individual cases.

At this time, two persons, Tomiyama and Hatsuki Tsujimura, who had begun working at the Laboratory a little earlier than Tomiyama on the recommendation of Yo Okada and was 13 years younger than Tomiyama, supported the Emperor's research. After Tomiyama passed away in 1981, Tsujimura was appointed as the director of the Laboratory (some documents describe him as a principal officer). After Tsujimura himself passed away, in 1989, immediately after the demise of Emperor Showa, no one was appointed to fill the position. Instead, Bungo Kawamura, a botanist, served the Emperor as a principal officer. Kawamura was promoted from being an officer at the Forestry Department, Minamitama Regional Office, Asakawa, Tokyo, to being an officer at the Imperial Household Agency on the recommendation of Kozo Hasegawa, a dendrologist, who served as the director of the Forestry Experiment Station. The samples from the laboratory, along with others, were moved to this station during World War II. Immediately after the war, the Emperor and the Empress began to collect plants at this station, where Hasegawa worked as the director, and thereafter, they continued to visit this place, for a total of ten times or so. Mayumi Yamada, a professor at Hokkaido University, and Shojiro Asahina, a renowned taxonomist of dragonflies, supported the Emperor in his researches on hydrozoans.

Now, I would like to devote some space to discussing the personality of Tomiyama, who was one of my former teachers, in more detail. Tomiyama, who was born in Busan, Republic of Korea, was educated at the Seventh Higher School under the old system (Zoshikan School) and then entered the Department of Zoology, University of Tokyo. He was outstandingly active in athletic sports in his high school days, and also in college, competing in the long jump with Mikio Oda, the first Japanese triple-jump gold medalist, whose name was bestowed upon an athletic field in Tokorozawa ("Mikio Oda Memorial Athletic Stadium"). Tomiyama initially set the record of 7 m 32 cm for the long jump. He told me that it was a new Japanese record at that time, even if only for a short period. Even with his relatively small stature, 167 cm in height, he displayed excellent physical performance. His manner of walking with a spring in his step was sufficiently picturesque for us to imagine his excellence as an athlete.

In his college days, he studied under the tutelage of Sigeho Tanaka, a pioneer in the field of fish taxonomy in Japan, and earned a doctorate degree for his classification of gobies. During the war, he was engaged in research at the Shanghai Natural Science Institute (presently under the umbrella of the Chinese Academy of Sciences), which had been established by the Empire of Japan in Shanghai in 1931, and, after the defeat in the war, was evacuated from Shanghai and worked as a teacher (professor) at Seikei High School under the old system. It was at this high school that I found myself under the tutelage of Tomiyama, who was teacher of our class. After a short time, he transferred to the Misaki Marine Biological Station, University of Tokyo, while he simultaneously supported the Emperor in his research on hydrozoans in the Biological Laboratory of the Imperial Household. After mandatory retirement from the University of Tokyo, he worked for a while at the Amakusa Marine Biological Station, Kyushu University, as the director. He was honest and rugged, and had so generous of a heart, partially because of his experience as an athlete, that he treated the entire staff equally, including the assistant collectors and manciples, and was known sometimes to dance the "Dojyo-sukui" (a dance representing the posture assumed while capturing weatherfish) when warming up at parties. Regardless of the fact that I was not a taxonomist, I worked at the Misaki Marine Biological Station as a research associate under his instruction for almost six years.

It is said that Emperor Showa seemed to have so outstanding a memory that he could keep his official duties, as well as any daily affairs, in mind with no need to take constant notes. Nagazumi states in his book [47]:

> To my knowledge, the Emperor has never taken notes then and there of whatever someone said. This means that the Emperor successfully managed his schedule according to his memory. Ichiro Tomiyama, an official in the service of the Emperor, said to me, "The Emperor's memory is far superior to mine!" The Emperor's mind is so admirable that he is able to keep track of what samples have been collected and when purely by memory.

There are many anecdotes about the Emperor that are widely known, including the fact that he was known to be able to remember individuals who were received in audience several years earlier. Such a capacity is ideal for taxonomists, who are required to have the ability to compare previous samples with new ones so as to identify both differences and similarities.

In the Biological Laboratory of the Imperial Household, the Emperor met with a great number of domestic and foreign biologists to discuss various themes with them, in addition to making researches. They later said that the Emperor was nothing but a true biologist in these instances. One of those biologists was Edred Corner, the first winner of the International Prize for Biology. Toku Tanaka states in his book, which describes the appearances of the rooms of the Obunko (meaning 'library') where the Emperor lived during the war, that, on the side desk, there was a pile of scientific journals and that original books on animals and plants occupied most of the four bookshelves. Also, there was a frame hanging on the wall in which a sketch of *Glyphus marsupialis* (*Sympasiphaea imperialis*), drawn using very realistic colors, was fitted. Although I will later describe the present conditions of the Biological Laboratory of the Imperial Household, it should be added here that "The Biological Laboratory of the Imperial Household" is generally simply referred to as "the Biological Laboratory," according to the intention of the present His Majesty the Emperor in 2008.

Collecting Marine Organisms in Sagami-Bay, a Worldwide Repository

Sagami Bay, where Emperor Showa collected a vast number of marine organisms, including hydrozoans, has been always known as the world's most renowned repository of these materials for specimens. This is because Japan was originally endowed with an abundant biota of marine organisms and, besides this, Sagami Bay has such a complex topography that a submarine canyon, called Sagami Trough, lies at a depth of over one thousand meters in its middle area.

A number of foreign teachers were invited to go there in the Meij Period, allowing them to get a head start on the researches of these marine organisms. Among this group, one pioneer was Franz Hilgendorf (1839–1904), a German teacher, who took his post in the preparatory course at Tokyo Medical School (the predecessor of the Faculty of Medicine, University of Tokyo) in 1873, and taught modern zoology and botany for its general education course for the first time in Japan. He, a man who visited fish markets every morning to collect fish and other marine organisms, is well known to have obtained Beyrich's slit shells (*Mikadotrochus beyrichii*), one kind of living fossil, and large beautiful spiral shells that lived in the deep seas, at a souvenir shop in Enoshima, Kanagawa.

Ludwig H. P. Döderlein (1855–1936) was also known to have collected fish at fish markets and many rare marine organisms in Sagami Bay, including the glass-rope sponge (*Hyalonema sieboldi*) of the class Hyalonematidae and the sea lily (*Metacrinus rotundus*) of the class Crinoidea. These specimens were collected in the part of the bay facing the area ranging from Enoshima to Misaki, and Döderlein reported this sea area to be a treasury of these marine organisms. These German teachers fostered no disciples specializing in zoology (as described later, Shin'nosuke Matsubara, a pioneer of ichthyology in Japan, did study under Döderlein); however, the Marine Biological Station, University of Tokyo, was established at Misaki on the suggestion of the latter teacher.

Whereas German teachers taught in the Faculty of Medicine, American professors did so in the Department of Zoology, Faculty of Science, University of Tokyo. Edward Sylvester Morse (1838–1925), well-known for his discovery of the Omori Shell Midden, was a disciple of Jean Louis Rodolphe Agassiz, who encouraged the development of zoology in the US. While visiting Japan to study lamp shells (Brachiopoda), including the genus *Lingula* in 1977, he was recruited to be the first professor in the Department of Zoology, University of Tokyo. He used a fisherman's hut situated at Enoshima as a marine biological station and collected a wide variety of lamp shells and organisms that lived on the rocky shore (Fig. 1.11).

The second professor in this department, who visited Japan two years later, was Charles Otis Whitman (1842–1910), although he himself did not do any collecting along the beach. In later years, he worked as the first director at the Marine Biological Laboratories (MBL) situated at Woods Hole, MA, in the US. Nevertheless, he eagerly offered his opinion on the need for a marine biological station at the university. Kakichi Mitsukuri (1857–1909), the third professor,

Fig. 1.11 Edward Sylvester
Morse, the first professor of
the Department of Zoology,
Faculty of Science, University
of Tokyo (Courtesy of the
University Museum,
University of Tokyo)

©The University Museum,The University of Tokyo

founded Japan's first marine biological station at Misaki in 1886 (Fig. 1.12). After
that, abundant animals and plants living in Sagami Bay were collected one after
another and their modes of life were revealed by the disciples who studied under
these foreign professors, the foreign researchers who stayed in Japan, and the
persons responsible for collecting organisms, including the afore-mentioned
Kuma-san. The investigation on marine organisms, including hydrozoans, from
Sagami Bay was conducted by Franz Theodor Doflein, as described earlier. But the
person most responsible for further enhancing the list of these organisms from
Sagami Bay was the Emperor Showa. Meanwhile, there are quite a few species that
still remain unknown living in the deep sea area of Sagami Bay (Cf. "The History of
Zoology in Japan" by Mohri and Yasugi [45] and "Fauna Sagamiana" [62]).

Using the Hayama Imperial Villa as a base of collection since before the war,
Emperor Showa had collected a vast variety of animals and plants from Sagami Bay
by means of: collecting on the rocky shores along Isshiki Beach, as well as the
Hatsuse and Aburatsubo Beach areas; dredging (trawling) the deep sea areas with a
collecting boat equipped with a dredge unit, on which the Emperor would ride; and

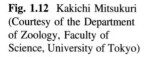

Fig. 1.12 Kakichi Mitsukuri
(Courtesy of the Department
of Zoology, Faculty of
Science, University of Tokyo)

sending divers into shallow sea areas. During his stay at the Hayama Imperial Villa, the Emperor concentrated on collecting organisms and sorting out his collection for as long as time would allow. In addition to his chamberlains and officials of the Imperial House, Empress Kojun and Princes and Princesses often accompanied him when he went out to collect organisms. When going to distant places on the large collecting boat "Hatagumo," the Emperor would take a utility boat from the pier of the Imperial Villa to Hatagumo, which was anchored in the offing. While seeing him off at the pier, Empress Kojun wrote a waka poem:

> Hoping the Emperor will find a variety of rare treasures, I see him on the boat parting from Hayama Beach.

He would collect for about two or three hours, two times a day, once in the morning and once in the afternoon. Collecting activities were frequently cancelled due to the weather conditions of the Hayama area, a lull in the wind in the morning or a strong wind in the afternoon. The details of the collecting activities and what the Emperor talked about were described in dozens of research notes written by the officials who served him, including Hirotaro Hattori, all of which have been archived carefully.

All of the Emperor's attendants said that he paid great attention to environmental preservation and species conservation in the places where he visited to engage in collection. To give an example, while collecting along rocky shores, he would turn over pebbles one by one to collect the animals and plants under them or attached to the bottom of them, and would then return the pebbles to their original positions. He asked others accompanying him to follow this practice. The Emperor also always asked his attendants to return specimens among the plants and animals scooped out of the sea by dredging, if there were any, that were determined to be unwanted during sorting out to their original sea areas. For this reason, the captains and persons responsible for collecting organisms had to keep note of the places where the organisms were collected. The Emperor even seemed to pay attention to which types of area, sandy or rocky, the collecting places were (quoted from the book by Tanaka [60]; "New Version of the Record of Experiences in the Imperial Court— Serving Emperor Showa" by Kinoshita [35] and "His Majesty the Emperor's researches in biology" by Habe [14]).

Now, we will explain dredging, one of collecting methods. A dredge is a tool for collecting the organisms that live at the bottom of the sea, kind of like a trawling net. The dredging method involves, for example, using a boat to pull a net stretched over a rectangular iron frame with teeth for fixation. However, while this shape of net could work successfully in sandy areas, it does not work in rocky areas. For this reason, to collect organisms in Sagami Bay, a special dredge, which was devised by Hachiro Saionji, an idea man, was used. This dredge had the structure of a 3 m-long log with a net attached to a chain on its back side, and an opening in the net, which was opened by means of glass balls and a wooden float. The log was intended to keep the net open and the chain would scuff along rocky surfaces (Fig. 1.13). This structure allowed a lot of animals to be gathered that could not be collected by a standard dredge. The animals and plants loaded onto the boat using the special dredge were sorted into individual wash basins in small quantities from buckets by the chamberlains and others, and then carried to the dock. The Emperor would carefully sort these animals and plants, put only those he determined to be essential into small glass bottles and take them with him to the Imperial Villa. Often, strenuous work on a boat may induce sea sickness, but the Emperor was resistant to this symptom. According to a well-informed source, when the Emperor visited Europe on the battleship "Katori" in his Crown Prince days, the crew members exclaimed, "His Imperial Highness was resistant to sea sickness" (quoted from the book by Hatano [17]).

The officials who served the Crown Prince were willing to support him to the utmost in his organism collecting activities, from those who worked with him prewar, as seen in the activities of the chamberlains Saionji and Kanroji, and those who worked with him after the war, including grand chamberlain Sukemasa Irie, deputy grand chamberlain Yoshihiro Tokugawa (the later grand chamberlain) and Hatsuki Tsujimura of the Biological Laboratory of the Imperial Household. All of these men played active roles in collecting organisms in Sagami Bay and plants at Nasu. Accordingly, they had naturally learned the names of the animals and plants themselves and called each other "Observers of the Imperial Court 'School,'" which

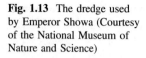

Fig. 1.13 The dredge used
by Emperor Showa (Courtesy
of the National Museum of
Nature and Science)

was derived from the Japanese saying, "The sparrows near the school sing the lessons." Yoshihiro Tokugawa also states, "His Majesty the Emperor and I had talks with each other, mainly about plants and animals," in "Grand Chamberlain's Will – Serving Emperor Showa for 50 years" [65]. The "Observers of Imperial Court 'School'" were more directly involved in the investigation into plants at Nasu, as described later.

In the following paragraph, some of the rare animals collected from Sagami Bay by Emperor Showa will be introduced by reference to "Those Who Came and Went from the Misaki Marine Biological Station," written by Isono [27]. One of these rare animals is *Atubaria heterolopha* (Japanese name 'Enokoro Fusakatsugi,' which means "covered with a tussock of tentacular arms"). *A. heterolopha* is, at most, a few millimeters in size and lives in deep sea regions. This animal is classified in the phylum Hemichordata, close to Vertebrata, together with acorn worms tens of centimeters in size, which live in the sand and mud areas of deep sea regions. This is because both species of organism have a structure that is the precursor to a notochord generated on the back side of the larva of a vertebrate; moreover, gill slits are found in both species. Furthermore, molecular phylogenetic analysis has confirmed that they are close to vertebrates. Acorn worms (Japanese name 'Giboshi-mushi' derived from an ornamental railing top, which resembles the shape of the animal's proboscis) are categorized in the class Enteropneusta, while *A. heterolopha* is in the class Pterobranchia. The latter organisms are quite rare, and only just over a dozen species have come to be known in the world. In waters close to Japan, only one sample of *Cephalodiscus levinseni* of the class Pterobranchia has been recorded, offshore of the Goto Islands, collected using a dredge by a Danish submarine cable layer.

In August 1935, Emperor Showa sent a dredge down to the sea bottom at a depth of almost 300 m on the west side of Jogashima Island to collect hydrozoans. At that time, as many as 43 chocolate-colored organisms about 1.25 mm in size, attached to the surface of one of the hydrozoans, were pulled up. The Emperor seemed to make a decision intuitively, when observing the specimens with Hiroo Sanada, an official serving as his right-hand man during the research, that those taken to the Laboratory with him might be of the class Pterobranchia. Moreover, Tadao Sato, a temporary officer who had been one of the Emperor's fellow students, continued his specialized research on them and revealed that they were a new species of a new class. He reported this new species in a German scientific journal by permission of the Emperor and gave them the Japanese name 'Enokoro Fusakatsugi' for *Atubaria heterolopha*. Unlike other species growing in clusters within a single sheath, they wound separately around the hydrozoan. *A. heterolopha* is one species of one class discovered so far and, since then, no record of this species having been collected has been found. Recently, however, just over a dozen species related to Fusakatsugi have been collected in the waters close to Japan, and Teruaki Nishikawa and Hiroshi Namikawa, both phylogenetic taxonomists, are conducting an analysis of these allied species, including their relationship with *A. heterolopha*.

In practice, the Emperor had always entrusted the academic presentations on the animals and plants, other than hydrozoans, in which he specialized, whether he collected them himself or not, to the specialists in the individual areas. This practice seems to resemble that of a patron, as seen in the European royalty and titled nobility, but has just as much to do with the position of a taxonomist, who carefully identifies species. When I participated in the practice of botany while collecting seaweed along the rocky shore of Misaki in my youth, Yukio Yamada, a professor at Hokkaido University and an authority on phycology, never immediately provided answers to questions from the students about the names of species; instead, he would later examine them and tell the students the correct names. I was very impressed by his sincere scholarly attitude. For readers' information, Yukio Yamada is the father of Mayumi Yamada, a specialist in hydrozoans.

By the way, in practice, scientists specializing in taxonomy (taxonomists) have almost no chance of collecting these biological materials by themselves; further-more, it is impossible to publish a book about the results of their studies unless they can afford to do so. The former grand chamberlain Sukemasa Irie states in the book "Observers of the Imperial Court 'School'" [24] that a publisher, who had known that the collector was the Emperor, was willing to make an offer to publish a book about the results of his study, which served as a starting point for professional scholars to enjoy chances to conduct their studies and gain support for their publication.

Another notable discovery by the Emperor was Koto-kurage (*Lyrocteis imperatoris*). In January 1941, just before the outbreak of World War II, the Emperor put a dredge into the sea at the point, commonly called "Hiramon," on the west coast of Enoshima, 70 m in depth. He intended to collect the root section of Torinoashi (*Metacrinus rotundus*), a primitive member of the phylum Echinodermata, of which a request for collection had been made earlier by the temporary officer Sato. However, they failed to collect the intended organism and, instead, two blob pieces congealed into jelly, which resembled jellyfish in appearance, were obtained. When the two blobs were attached to each other to restore their original shape, it became evident that the intact individual was a U-shaped organism 10 cm in size, 30 cm in total size when its tentacles were included. Several days later, and then again in July and November, nearly intact individuals were collected with dredges in the same area, finally reaching a dozen in total. Taku Komai, to whom the Emperor entrusted the research on identification of the organism, made it clear that they might be classified into a new species of comb jellyfish, and gave this species a name, "Koto-kurage (*Lyrocteis imperatoris*)." As mentioned earlier, whereas jellyfish belong to the phylum Cnidaria (Coelenterata), despite being called "jellyfish," the comb jellyfish belongs to another phylum, Ctenophora, and has arrays of cilia, called comb plates, on its body's surface. Recently, we have gotten many chances to see comb jellyfish in images from the deep seas and other places. Furthermore, small larvae were born from one of the collected samples of *L. imperatoris* being cultured, which revealed its viviparous mode of birth and defined its developmental stages.

In fact, Alan Owston, an English trader who resided in Yokohama, collected one portion of an unknown organism from a shallow in the offshore of Sagami Bay using his yacht in the summer of 1896. Tokichi Nishikawa described the unknown organism as a "strange animal" in "Dobutsugaku Zasshi." (Some say that Nishikawa collected Koto-kurage using Owston's yacht.) Its true identity was revealed later; it was simply a Koto-kurage. As an aside, Tokichi Nishikawa, who graduated from the Department of Zoology, Faculty of Science, University of Tokyo, and married a daughter of Kokichi Mikimoto, the founder of the Mikimoto Pearl Co., succeeded in the very first aquaculture of perfectly round pearls. Toku Tanaka describes, in "Emperor Showa and His Biological Research" [60], that Emperor Showa and Hiroo Sanada thought that the "strange animal" might be a "Koto-kurage" when they saw the collected animal for the first time, and that the Emperor himself determined that it produces larvae instead of eggs and spermatozoa.

Moreover, both jellyfish, Hinomaru-kurage (*Steleophysema aurophora*), which is a rare species of Siphonophora collected offshore of Hayama Beach in 1935 belonging to the same class as that of *Physalia physalis*, and Nokishinobu-kurage (*Athorybia rosacea*), collected in the same area in 1950, were recorded together as new species by Tamiji Kawamura, a professor at Kyoto University. Around the same time, Taku Komai made it clear that *Stephanoscyphus coniformis*, a member of the class Hydrozoa collected offshore of Arasaki Beach near Hayama, was different from that collected in the sea area near the Seto Marine Biological Station,

Kyoto University, situated at Shirahama, Wakayama. Only two species of *Stephanoscyphus* have been found in Japan. The Emperor also found, for the first time in Japan, a species of boxer crab in each of whose claws was pinched a small sea anemone (*Lybia tessellatae*), offshore of Hayama Beach at a depth of 83 m in 1957. It is said that, when being attacked by an enemy such as a devil fish, a boxer crab gets away from them by brandishing sea anemones over its head pinched in its claws. The reason that it does this is that these enemies will retreat for fear of poison being injected into them from the nematocysts through the tentacles. This species of crab was called "Hatagumo-gani (*Lybia hatagumoana*)," after the name of the collecting boat used at that time by Tsune Sakai, a professor at Yokohama National University, who was well-known as an expert on crabs. Hozuki Pinno, one species of the Kakure-gani (*Geograpsus grayi*), which is parasitic and lives in Hozuki-gai (lamp shells), belonging to the class Brachiopoda, is also a rare organism. These animals are shown in Fig. 1.14.

As previously described, Okinaebisu (pleurotomariid slit shells), one species of living fossil belonging to the class Gastropoda, can be collected in Sagami Bay. The Emperor collected a beautiful Beni-okinaebisu (*Mikadotrochus hirasei*), which had just died, on the beach at the beginning of the Showa Period, and was fascinated by the modes of life concerning this species of organism. Taking the Emperor's interest into consideration, Hachiro Saionji went to the area around the Uraga Channel, where the Imperial Japanese Navy was strictly controlling access so as to

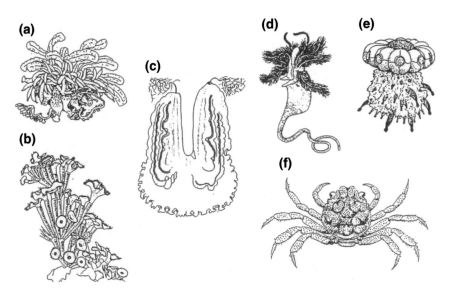

Fig. 1.14 Marine organisms discovered by Emperor Showa: **a** *Athorybia rosacea* (Nokishinobu-kurage), **b** *Stephanoscyphus corniformis* (Iramo), **c** *Lyrocteis imperatoris* (Koto-kurage), **d** *Atubaria heterolopha* (Enokorofusakatsugi), **e** *Steleophysema aurophora* (Hinomaru-kurage), **f** *Lybia hatagumoana* (Hatagumo-gani) (From the book "Persons Who Came and Went to and from the Misaki Marine Biological Station," written by Isono [27], Courtesy of Yuko Isono)

Fig. 1.15 Emperor Showa collecting specimens along the Hayama Coast, with Empress Kojun, in 1961 (Courtesy of the Imperial Household Agency)

guard the secrecy of warships, and finally obtained living samples using an improved dredge of his own design. They say that the Emperor was very pleased. The Emperor also made a study on Ginebisu (silver margarite, *Lischkeia argenteonitens*), unique to Japan, belonging to the same class of Gastropoda as Okinaebisu, which emits a pearl color. A new species of Ebisu-gai, which is closely related to *L. argenteonitens*, was also collected along with the latter. The Emperor requested that Tokubei Kuroda, an expert from Kyoto University, identify the new species. A scientific name, 'Tsuchiya Ebisu,' was given to it, after chamberlain Tsuchiya, who had served the Emperor from his childhood and contributed a great deal to guiding the Emperor towards his course of study in biology.

The Emperor wrote the following waka poem in honor of the pleasure he had gotten from resuming his habit of sample collection on the beach after the end of the war (Fig. 1.15):

> Pushing through seaweed swinging among rocks at low tide, I collect sea slugs from beneath a rock under the blazing sun.
>
> Digging in the sand of a calm mudflat at low tide, I've finally found an Omidori Yumushi (spoon worm, *Urechis unicinctus*).

Sea slugs with a degenerated shell belong to Opisthobranchia, the class Gastropoda, the phylum Mollusca, many of which exhibit a variety of bright colors. They are relatively close to the familiar sea angels (*Clione*) that live in the northern seas. Spoon worms (*U. unicinctus*) of the phylum Annelida are closely related to bristle worms (Nereidae) and earth worms. They are used for fishing lures, as well as for food in some regions.

Emperor Showa's Collecting Boats, Then and Now

Everybody knows that a boat is indispensable for collecting marine organisms in sea areas far from the beach. According to one explanation, Hattori advised the Emperor to make a study on marine organisms so as to be freed from the uncomfortable feeling resulting from the tight security he was forced to endure, because only a limited number of attendants, including chamberlains, a boat man, and the persons responsible for sample collection, would be on the boat. Hatsuki Tsujimura, a chamberlain who served the Emperor from the end of the war, describes, in "Observers of the Imperial Court 'School,'" [24] that a black board with boat names written on it had been left in a working cabin at Hayama since the prewar time. The names written on the black board included the "Hayama-maru," the "Miura-maru" and the "Najima-maru." It is assumed that the "Hayama-maru" was an Imperial boat (gozabune) for collecting samples, the "Miura-maru" was used for dredging, and the "Najima-maru" carried divers. However, these boats were bestowed upon the Imperial Navy during World War II. In the spring of 1948, soon after Tsujimura had arrived at a new post, the Emperor visited the Hayama Imperial Villa for the first time after the end of the war. In fact, according to a well-placed source, the Emperor had his first chance to visit Hayama in spring, when the tidal current was best suited for sample collection on the beach. At that time, a Japanese-style boat, called "Take" ("Bamboo") was used as the sample-collecting boat. Take, which had been used before World War II, was a wooden sculling boat with 1.5 t in displacement capacity and was steered by two boatmen, Yugoro Moriya and Zentaro Tsunoda. Another Japanese-style boat, "Momo" ("Peach"), was also used as a sample-collecting boat. Take was later given to a person who liked to fish. In the summer of the same year, the Emperor took on a challenge to make "Hayabusa Go," a motor boat, tow the Japanese-style boat for dredging. Unfortunately, as a matter of course, the sample collecting area was too narrow to achieve full-scale dredging.

To solve this problem, a new sample-collecting boat was built in 1949 and was named the "Hatsuse-maru." Hatsuse is the name of the place on the south side of Hayama where the Imperial Household Agency had made an attempt to build an Imperial Villa because of its convenience for sample collection. Although the "Hatsuse-maru" was a wooden service boat equipped with a winch for rolling a dredging seine onto a reel, it was almost the same size as "Take" and the conventional fishing boats commonly found here and there. Ultimately, however, this boat was only allowed to engage in sample collection on the coast, because of its poor body balancing and the winch's failure to work in a normal manner. Moreover, the whereabouts of the "Hayama-maru," which, as mentioned above, had been bestowed to the Imperial Navy, had been determined, and it was decided that the boat would be returned to the Imperial Household Agency. For this reason, the "Hatsuse-maru," being no longer in use, was bestowed to the Misaki Marine Biological Station, University of Tokyo, probably in 1955. This fact may have influenced the favorable treatment that Ichiro Tomiyama, the director of the Station, enjoyed after he started working at the Biological Laboratory of the Imperial Household. At that time, the "Rinyo-maru," the boat then in use at the Misaki Station, named after Rinkaijikkenjo (Marine Station) and Yo K. Okada, the former director of the Station, was beginning to become too old to use. This was the exact time when I arrived at my new post, as a research associate (currently known as an assistant professor). The "Hatsuse-maru" contributed to the training of students and the collection of materials for researchers, including my own, but with the construction of a new boat, it was left on land. Among the imperial collecting boats used by Emperor Showa, the "Hatsuse-maru" fell to a somewhat unfortunate fate, most likely being incinerated around 1974 due to its age. It did not have a chance to be used by the Emperor to sail oceans far away from the Hayama Imperial Villa, though the researchers, including myself, often used it to go to Tokyo Bay to collect samples [44].

Tracing back the history of the "Hayama-maru," it was built at the former Naval Arsenal situated at Yokosuka as a sample-collecting boat in 1934. The boat, which had 16 tons in displacement capacity and was equipped with a winch for dredging, had played an important role during the period up to 1941, when the Pacific War broke out. During the war, it was used for training at a Naval Academy in Edajima and, after the war, it was requisitioned once by the Occupation Army and then used as a patrol boat by the Japanese Coast Guard. After returning to its original place, the Hayama Imperial Villa, in 1950, the "Hayama-maru" was again used to go to deep sea areas that the "Hatsuse-maru" had never been capable of reaching, and contributed to collecting a wide variety of marine organisms up to 1956. The Emperor visited the Misaki Marine Biological Station on the "Hayama-maru." At present, it is displayed at the Omishima Maritime Museum at the Oyamazumi Shrine situated on Omishima, the Seto Inland Sea, as a result of the fact that various kinds of tangible movable property, including this boat, owned by the Naval Academy had been entrusted to the shrine to prevent them from being condemned by the Occupation Army. The Municipal Government of Hayama, where the Imperial Villa is situated, hopes someday to have it returned to their town, but this has not yet been realized (Fig. 1.16).

A successor to the "Hayama-maru" was the "Hatagumo," with 45 tons in displacement capacity. It was used exclusively by the Emperor as a boat for collecting marine organisms, 52 times in total during the period from 1956 to 1971. It was said that six crew members, the Emperor and ten of his attendants were on board. Going by my own memory, the "Hatagumo" had been a patrol boat owned by the Japanese Coast Guard that was eventually returned back to its original duty; at present, it is archived at the Japanese Coast Guard Academy, Maizuru, Kyoto. Another successor boat to the "Hatagumo" was the "Matsunami," a patrol boat with 83 tons in displacement. With the construction of the Suzaki Imperial Villa in 1971, the

Fig. 1.16 The Imperial collecting boats: the "Hayama-maru" (upper) and the "Hatagumo" (lower) (Courtesy of the Imperial Household Agency)

locations for sample collecting changed from those offshore of Hayama to Suzaki offshore the Izu Peninsula. Just around that time, however, the first oil crisis occurred. For this reason, taking into consideration that any waste of fuel by the collecting boat and the patrol boat of the Japan Coast Guard that escorted it needed to be avoided, the Emperor suspended use of the collecting boat, choosing to engage in sample collection only with divers in limited regions of the sea starting around 1973. The "Matsunami" was in commission until 1995. The current patrol boat, "the second Matsunami," a successor to the retired "Matsunami," with 204

Fig. 1.17 Emperor Akihito rowing a boat, accompanied by Empress Michiko, Princess Akishino and Prince Hisahito (Courtesy of the Asahi Shimbun Company)

tons in displacement capacity and VIP rooms for welcoming honored guests, also serves as a Royal Yacht.

Tracing a path through the first half of the Showa Period, including the period during the wars, sample collection was done using simple boats that were not much better than small-sized fishing boats, including Japanese-style sculling boats. They were far more modest than luxury royal yachts, as well as the marine research vessel owned by Albert I, the Prince of Monaco. In the second half of the Showa Period, the patrol boats also served as collecting boats, allowing for trips out onto the broad ocean, but the displacement capacity of the largest one among these boats was, at most, 83 tons. For this reason, taking the Emperor's intention into account, the locations for sample collecting were limited to the sea regions around Sagami Bay, offshore the coasts of Hayama and Suzaki. By the way, it is widely known that the present Emperor Akihito also sculls a Japanese-style sculling boat on his own, with his children and grandchildren on board, along the sea coasts, including that of Hayama (Fig. 1.17). Only people with experience will recognise how difficult it is to skillfully manage a Japanese-style boat while pulling an oar without dislocating it from the oar lock. Nevertheless, unlike Emperor Showa, Emperor Akihito has never used a collecting boat for his own exclusive use.

Animals and Plants Living in the Imperial Palace and Annual National Tree-Planting Ceremonies

As will be described later, Fukiage Garden, where Obunko (an air-raid shelter at the Imperial Headquarters constructed during World War II) and the Fukiage Imperial Palace (in which the Emperor and the Empress lived) were situated, was a magnificent garden of the Edo Period. This notable garden, which became desolated at the end of the Edo Period, was restored under the instruction of Emperor Meiji. Moreover, in this garden, a nine-hole golf course had been already built when Emperor Showa returned from Europe in his Crown Prince days. The area around the sites of the Biological Laboratory, the Farm Field, and the Imperial Household seems to have once been used as a riding ground. The Emperor discontinued playing golf when the Sino-Japanese war broke out. Saying "I prefer a natural field with native grass growing to a garden with a well-kept green," he decided to convert the garden into a natural field (quoted from the afore-mentioned book by Tanaka [60]). His mother, Empress Dowager Teimei, agreed to the conversion of the garden to the natural field, possibly because it was likely to be a place full of memories of Emperor Taisho for them. By that time, Fukiage Garden had long been neglected, and Japanese silver grass (*Miscanthus sinensis*), bellflower (*Campanula punctata*), fringed pink (*Dianthus superbus var. longicalycinus*), and other kinds of wild grass had grown in thickly. Some of the officers of the Imperial Household Agency stated opposition to the conversion, because it had once been such a great garden, and the gardeners, who were unable to stand by with folded arms, neatly cut the tufts of grass so as to fulfill their duty. This caused the Emperor a great deal of disappointment. After the war, Ryogo Urabe, who had served as a chamberlain since 1969, made the same mistake shortly after he was appointed to the new post. According to reliable sources of information, the Emperor confided to one of his attendants:

> I cannot manage anything, except this inner court, at my own discretion. Well, let us leave it as a wild-grass field (Kanroji [33]).

This was the genuine intention of the Emperor, who was the Generalissimo of the Imperial Japanese Army and Navy and had been ranked as the richest person in the world in the past because all of the nationally owned lands were regarded as his property. The Emperor cultivated a new hobby in which he always brought an illustrated reference guide of plants, instead of his golf clubs, with him when walking outside with the Empress and his chamberlains. The Emperor continued this hobby throughout his life. The Empress always accompanied the Emperor, but it became more and more difficult for her to walk due to a lumbar spine fracture she had suffered at Nasu in 1977 and a fall she had taken in the Imperial Palace at the end of 1984. Finally, regrettably, the Emperor had to go out alone later in life.

Meanwhile, the gold standard of illustrated reference guides to plants was "Makino's Illustrated Reference Guide of Japanese Flora," written by Tomitaro Makino (1862–1957). Makino, who was born in Kochi Prefecture, had devoted himself to botanical study on his own in his childhood and joined the researches at the Department of Botany, University of Tokyo, with the permission of Ryokichi Yatabe, the first professor at the Department. Makino was later appointed as a lecturer and worked at the University for 47 years. Having given names to about 2,500 species, he is worthy of being called the father of plant taxonomy in Japan. He was given the first "Person of Cultural Merit" award, and, after he died, the "Order of Culture" was bestowed upon him for his glorious achievement. He passed away in 1957 at the age of 95.

In the fall of 1948, Makino, at the age of 86, visited the Imperial Palace by invitation. According to the descriptions in the book by Tanaka [60], Japanese silver grass (*Miscanthus sinensis*) made the native grass garden of Fukiage Palace particularly attractive. Having awaited Makino's arrival for a few hours, the Emperor asked him a question as soon as he arrived at the Imperial Palace: "In regard to the Syajikumo (stonewort, *Chara braunii*) over there. What is its nature?" In answer to the Emperor's question, Makino said, "The Syajikumo growing in this old pond was discovered by a person working at the Natural History Agency in 1871, and he is the one who gave it this name. Syajikumo may be classified into three different species, but it is unknown what species the Syajikumo found here belongs to, because no comparative sample remains anywhere." Syajikumo, a species of green freshwater alga, is apparently rather similar to field horsetail (*Equisetum arvense*) and has very large internode cells, making it easy to observe protoplasmic streaming. The two men walked through the garden, the Emperor asking Makino various questions about the plants they observed along the way. The Emperor expressed his appreciation for Makino's knowledge, and, when they reached the edge of the garden, he said to him, "Are you tired? You should take care of yourself." After that, the Emperor went to the Biological Laboratory of the Imperial Household. Makino, before returning home, promised the accompanying chamberlain, "If I obtain any plants from Musashino that are not found in the Fukiage Garden, I will certainly bring them with me to the Biological Laboratory of the Imperial Household" (quoted from the book by Tanaka [60]). The Emperor made his taxonomical study (regarding the genus and species) of the plant specimens that he had collected under the tutelage of Makino.

A previous attempt had been made to transplant natural plants growing at Musashino into Fukiage Garden. The chamberlains, including Irie, Tastuo Sato, who was known to be an amateur botanist and had served as the Director-General of the Cabinet Legislation Bureau and the President of National Personnel Authority, and others, brought native grasses and trees to Fukiage Garden from the current Institute of Natural Study in Shiroganedai, Minato-ku, Tokyo, the imperial estate at Kitami, Setagayaku, Tokyo, and fields and mountains such as Mt. Takao, Asakawa, and Totsuka. Totsuka is the place where Busei Yoshida, who worked at the Imperial Household Agency for three generations of Emperor, Meiji, Taisho and Showa, resided ("Kyuchu Jijyu Monogatari" ("The Story of the Court Chamberlains") by Irie [25]). These transplanted plants formed a native grass garden that bore a close resemblance to unspoiled Musashino, together with the grasses and trees that grew naturally because no mowing had been done. Whenever the chamberlains used the word "weeds," the Emperor would reprove them for their thoughtlessness. "No plant should be called a weed." For this reason, the word "weed" was taboo among them (Kageyama [31]). Needless to say, there was a lot of knowledge on nature and plants to be gained in Fukiage Garden and the Nasu Highlands.

In addition to Tatsuo Sato, the Emperor was also served by the academic consultation of the Goninshu (a five-botanist group) made up of Masaji Honda and Hiroshi Hara from the University of Tokyo, Shiro Kitamura from Kyoto University, Arika Kimura from Tohoku University, and Hiroshi Ito from the Tokyo University of Education (the current University of Tsukuba). There was also Yoshio Kobayashi from the National Museum of Nature and Science, a researcher on mushrooms, and Masami Sato from Yamagata University, an authority on lichens who had graduated from the University of Tokyo. As he worked at Yamagata University, the chamberlains were hesitant to ask him to serve the Emperor as an academic consultant, since Yamagata University was not a former imperial university. The Emperor said:

It is far preferable to invite a person who works at a local university as far as circumstances permit (Irie Sukemasa's Diary [26]).

In fact, the number of taxonomists had not only decreased at the Departments of Zoology at seven old imperial universities (Tokyo, Kyoto, Tohoku, Kyushu, Hokkaido, Osaka, and Nagoya), but in their Departments of Botany as well. Roughly ten of my classmates who graduated from the Department of Botany, University of Tokyo, in 1953, when Honda and Hara taught there, joined the physiological or biochemical laboratories, and none of them entered the taxonomical field. For reader's reference, as many as 1,470 species of plants are described in the last Emperor's book, "Plants of the Imperial Palace."

As the foliage gets thicker, the number of animals living within it increases. It has long been said that raccoon dogs might live on the grounds of the Imperial Palace and, even now, several generations make their home there, as will be described later. In addition, many kinds of bird, including pheasants (*Phasianus versicolor*), mandarin ducks (*Aix galericulata*), and spot-billed ducks (*Anas zonorhyncha*), live there or stop there to rest during migration. In 1967, when the Empress spotted a stray hoopoe (*Upupa epop*) flying directly opposite to the dining hall, the Emperor, who was familiar with this kind of bird because he had seen one while touring Manchuria before the war, said, "That's a hoopoe!" (Tokugawa [65]). The hoopoe is similar to the thrush (*Turdus naumanni*) in size, and its head crest sometimes features ruffles. The errant hoopoe stayed around the Imperial Palace for a while, and the Empress wrote many waka poems about this bird and drew a picture scroll of it.

In the past, fireflies have lived on the grounds of the Imperial Palace, but they disappeared at some point. A great many Genji-botarus (*Luciola cruciata*) from various areas in Japan were presented and released, but a population failed to develop. During the war, in 1944, after black snails (*Semisulcospina libertina*) had been released, Heike-botarus (*Luciola lateralis*) emerged, instead of Genji-botarus. Most probably, the offspring of the Heike-botarus had been living hidden somewhere. Since the recent document mentions the name of the Genji-boraru, it is believed that those, having been brought to the Imperial Palace by someone, have successfully propagated. Moreover, the Imperial Palace is also a comfortable permanent safe refuge area for edible frogs (*Rana catesbeiana*) and toads.

In front of the Biology Laboratory of the Imperial Household, there was an agricultural field, in the corner of which a rice paddy was formed. Indeed, it seems that the rice field had already been formed in the Crown Prince's Palace, Akasaka. The Emperor started planting and harvesting rice there in 1929, at the glowing suggestion of Yahachi Kawai, the grand chamberlain at that time, who desired that the Emperor gain a deep understanding of agricultural practice. In the following year, the Emperor Showa began offering newly-harvested rice to all of the deities at the Niiname-sai Festival (a court ritual celebrating a good grain harvest) in November every year. This established a link between the Emperor's biological research and established rituals. As a counterpart to this linkage, the Empress took over the duties of Empress Dowager Shoken and her successors, that is, rearing silkworms and Japanese oak silkmoths (*Antheraea yamamai*) at the Sericultural Station situated near Momijiyama. The Emperor himself planted three rice stubbles each in two places that year. Unfortunately, he died the very next year, and was therefore unable to fulfill his hope to plant rice the next year as well. This event has continued up to the present Emperor. According to informed sources, Emperor Showa also carried out experiments on rice hybridization.

The Emperor's intention to conserve nature was spread beyond the Imperial Palace throughout the country through the "National Tree-planting Ceremony." The "National Tree-planting Ceremony" is a core activity of the "National Tree-planting Campaign," which was commenced with the intention of recovering plants in a landscape devastated due to the war. The first event of the "National Tree-planting Campaign" was called the "Tree-planting Event and National Land Tree-planting Convention" in Onshi (given by the Emperor) Katayama Forest, Kofu, Yamanashi Prefecture, in 1950. Since the 21st convention in 1970, in Fukushima, this event has been called the "National Tree-planting Ceremony." As an annual event, the convention is always held with their Majesties in attendance. The Emperor makes a speech, and their Majesties themselves plant trees and tree seeds. The convention already travels through all of the prefectures in the country and is now entering its second round. The trees planted in past conventions have often grown up to form fine, flourishing forests. Emperor Showa himself is said to have believed that a broadleaf forest would be more beneficial for water control than a coniferous forest, which is now accepted knowledge.

Some criticize the "National Tree-planting Ceremony" for the reason that it destroys the natural environment, rather than conserving it, through actions such as venue preparation and road work. This criticism should be directed toward the local government as the host, because the Emperor, who truly loved nature, absolutely did not intend for such destruction of nature to take place. Nevertheless, it has been said since old times that, "The places where the Imperial Family members visit are neatly maintained." The present Emperor Akihito, who is also extraordinarily interested in plants, stated at the 2003 "Tree-planting Ceremony":

In Japan, the forests have been brought to ruin due to excessive logging for the purpose of meeting the great demand for timber during and after the war, which often caused serious disasters, such as great typhoons and heavy rains; however, it is reassuring to me, even under such a situation, that a wide range of intensive efforts made by the public towards tree-planting have borne fruit, with forests useful for peoples' lives being grown with a wide variety of aspects in mind, including the use of timber, prevention against disasters and the recharging of water resources.

Meanwhile, in recent years, continuing depopulation in mountain villages and an aging population in the forestry industry have made it extremely difficult to take good care of the forests so that they might continue to be revitalized. In this context, it is very pleasing that forest creation activities have recently being advanced by participants from a variety of sections of the public. However, it is regrettable that the "speech" by the Emperor, according to custom, has been discontinued so as to reduce the Emperor's official duties.

Botanical Collection at Nasu and Suzaki

As described earlier, the Nasu Imperial Villa was built for Emperor Showa and Empress Kojun, a newly-wed couple at that time, on the hillside of the Nasu Highlands. The royal couple visited there almost every summer, from the year when the Imperial Villa had been built to the year before Emperor Showa's demise, the only interruption being during the war. At the Nasu Imperial Villa, the Emperor initially made a research mainly on slime molds, and then changed his research subject to higher plants such as ferns and seed plants around 1953. At Nasu, he also often took walks with the Empress carrying a pictorial guide of Japanese flora and reference guide books. When venturing out to places far from the Imperial Villa, he would go on horseback in his early years and by jeep or the like later on.

The afore-mentioned Goninshu (five-botanist group) of Masaji Honda, Hiroshi Hara, Shiro Kitamura, Arika Kimura, and Hiroshi Ito, as well as Tomiyama, Tsujimura and Kawamura of the Biological Laboratory of the Imperial Household and others, visited Nasu to support the Emperor's research (Fig. 1.18). The rooms for these botanists were at the Imperial Villa. The Emperor, owing to his intention that not only the valuable species of marine animal collected in Sagami Bay be conserved, but the plants as well, checked the area around the observation spot to see if multiple organisms of the same species existed, so as to avoid more collection than necessary.

In addition to the above-mentioned botanists, the "Observers of the Imperial Court 'School,'" including Irie and Tokugawa, greatly contributed to the Emperor's observatory research. With something resembling a drawing board hanging around their necks, they recorded the plants' names and such detailed data as their identity and whether they were flower or fruit, all set down on the ruled paper that was for the Imperial Household Agency's exclusive use, as instructed by the Emperor. To avoid any delay in their investigation, by agreement, they did not express their

Fig. 1.18 Botanists and others working at the Imperial Household at the Nasu Imperial Villa in 1970 (From the leftmost: Hatsuki Tsujimura, Ichiro Tomiyama, Hiroshi Hara, Shiro Kitamura, Masatsugu Honda, Arika Kimura, Hiroshi Ito and Bungo Kawamura, courtesy of Itsuki Saito)

individual opinions at all. Bungo Kawamura discovered a fine iris growing natu-rally along a stream that ran around Hitotsu-momi at Nasu, and Hiroshi Hara gave it the name "Nasu Hiogi Ayame" ("*Iris setosa var. nasuensis*"). "Nasu Hiogi Ayame" is about 1 m in height and gives bloom to violet flowers larger than those of "Hiogi Ayame" ("*Iris setosa*"). At present, its colony in Tochigi is conserved as an endangered species and it has also been planted in the Nikko Tamozawa Imperial Villa Memorial Park and the Nikko Botanical Garden, University of Tokyo. Kawamura was exclusively engaged in the investigation of the plants around Nasu on behalf of the Emperor when the latter could not visit Nasu or could not reach his destination.

The Nasu region exhibits a wide variety of flora, particularly where the southern and northern zones of assemblage and those distributed along the coast of the Pacific

Ocean and the coast of the Japan Sea draw a boundary between each other or are mixed together. There grow Nasu-specific species such as *Salix xeriocataphylloides*, *Salix x. nasuensis*, *Eriocaulon nasuense* and *Persicaria amblyophylla*. As development advanced from year to year, new species were discovered, while some became extinct. Emperor Showa states, in the Introduction to "Flora Nasuensis":

> I feel a fresh breath of nature whenever I come up to Nasu. Communing with its nature, I have passed this last summer again and continued my observation of the plants, and have been able to add some more new facts to the contents of this book already prepared. I do wish all the wild animals and plants to be allowed forever to continue to live their own lives in quiet surroundings, and with this sentiment, what I mean by this publication is only to keep as exact records as possible of the aspects of lives of the plants at Nasu.

"Flora Nasuensis," issued in 1962, which the Emperor had been involved in reviewing since the first proof stage, describes 78 species of fern and 1,078 species of seed plant. Subsequently, his research advanced even further, and "Flora Nasuensis: Additions and Emendations" and "Nova Flora Nasuensis" were issued. "Nova Flora Nasuensis Supplementum," issued in 1985, records 143 species of fern and 1,828 species of seed plant. Thus, the Emperor recorded in detail the changes in vegetation in the places that he had chosen for observation. Although they are low-profile works, needless to say, they are extremely valuable materials.

At Nasu, pheasants and copper pheasants have often been released by the Emperor. This event has been continuously observed, even up to the present.

The Suzaki Imperial Villa, situated at Shimoda on the Izu Peninsula, was constructed in 1971 as a successor to the Numazu Imperial Villa. The "Marine Biological Laboratory" was attached to the new Imperial Villa. It was decided that the base ground for research on marine organisms would be transferred to Suzaki, and an investigation was carried out in the sea areas along the coast of Suzaki using the collecting boat the "Matsunami," partially because the main building of the Hayama Imperial Villa had been burnt down in an arson attack early in 1971. However, they were only able to carry out their collecting activities using the boat there for a period of two or three years, due to an oil crisis and for other reasons. Since then, marine organisms have been collected only on the rocky shore or by skin divers.

In contrast, an investigation into plants has been actively conducted since then. As described earlier, Emperor Showa, Empress Kojun, their chamberlains, the member of the Biological Laboratory of Imperial Household, Tatsuo Sato, and the five-botanist group would, as usual, visit the "Marine Biological Laboratory" at the Suzaki Imperial Villa to engage in investigations during short periods throughout the four seasons. Izu Peninsula, on which Suzaki is situated, belongs to the rift valley, the Fossa Magna zone, on the east side of the line between Itoigawa and Shizuoka, and produces plant species specific to this region. According to the description in "Flora Suzakienesis," written by the Emperor:

> Suzaki and its neighbourhood, that is a part of Southern Izu, are influenced by a warm current, and therefore have rich growth of warm-region (subtropical) and coast vegetation. Naturally, these plants are different from those at Nasu, and are similar in many respects to plants at Ashizuri Promontory in Kochi Prefecture [46].

Some plants, such as *Heterotropa muramatsui* and *Allium schoenoprasum* var. *idzuense*, are limited in their distribution to this area. Many ferns found here are evergreen and characterized as forming northern-limit distributions. "Flora Suzakienesis" records 62 species of fern and 535 species of seed plant. This area is known for the fact that Matthew C. Perry, when he visited Shimoda while leading Kurobunes (black warships), made his attendants collect the plants growing around there. The Emperor wrote a waka poem at Utakai Hajime (a poetry reading held at the beginning of the New Year) with "Midori (green)" given as a theme in 1984:

> A thick layer of green usuba-aonori (*E. linza*) covers the surface of a rock left standing on Suzaki Hama (beach) at ebb tide.

Emperor Showa's Second Visit to Europe and First Visit to the United States

In consideration of the fact that fifty years had passed since Emperor Showa visited Europe in 1921 when he was the Crown Prince, Eisaku Sato, the prime minister at that time and a Nobel Peace Prize laureate, devised a plan in 1970 that the Emperor, together with the Empress, would revisit Europe. Based on the plan, in 1971, the Emperor actually did travel abroad, according to a schedule by which he would depart on September 27 and return on October 4. This time, the Emperor visited Europe by airplane instead of warship. The visiting countries included Denmark, Belgium, France, the UK, the Netherlands, Switzerland, and West Germany. Among these, the visits to three of the countries, Belgium, the UK and West Germany, were formal. At that time, no airplanes were allowed to fly over the USSR, and, when following the North Pole Route, they had to fly via Anchorage International Airport, a transfer station. When I visited Europe, I often stopped over at that airport, which included a building with a big stuffed polar bear exhibited in it. When the Emperor and the Empress arrived, President Richard Nixon and First Lady Pat Nixon were there to welcome them. This was the first time that the Emperor had stopped over in a place in the US. At that time, the US had sunk into the morass of the Vietnam War. The Emperor used the opportunity to express his thanks to the President for the assistance provided by the US government after the war, even though the US was under such a dire situation.

After leaving Anchorage, the Royal couple took a quick side trip to Copenhagen, Denmark, and then visited Belgium, the first of the official visits. In Brussels, they attended a banquet hosted by King Baudouin, and the Emperor made a speech stating that Japan longed for peace. In the UK, at a banquet hosted by Queen Elizabeth at Buckingham Palace, he spoke of his wonderful memories from when he had visited the country in his Crown Prince days:

> Since we arrived at the airport, I have been feeling the same warmth in your people's hearts as I enjoyed fifty years ago. I have always held close the thoughtful words spoken to me by His Majesty George V at the time, as if he was my own father.

Silk fabrics woven using silk threads spun from silkworm cocoons reared by the Empress in the Momiji-yama Sericultural Station of the Imperial Household and paintings drawn by the Empress herself were presented to the royal families, who were pleased with these gifts. The Emperor met with the Duke of Windsor (Crown Prince Edward at that time), who had lived in Paris since his abdication, also for the first time in about fifty years. They were welcomed in all of the places that they visited, thus achieving the fruitful results of Imperial diplomacy. On the other hand, there were protests in some countries, partially due to the impact of World War II, with incidents such as bottles filled with water being thrown at the motorcade in which the Emperor and Empress were riding. It is said that they encountered the most violent demonstration while visiting a zoo in Amsterdam, since the Netherlands had experienced the loss of a colony in Indonesia and had had many of its people there taken prisoners due to the war. The demonstration was so violent that Irie, the then-Grand Chamberlain, wrote in his diary, "I felt relieved when we left the Netherlands" [26].

Even on days during his visit to Europe when he was busy with official duties, the Emperor found time to discuss hydrozoans with local experts. He had conversations with Professors Paul Crump and Amanuensis Petersen, who were both biologists specializing in hydrozoans in Denmark; Professor Eugene Leloup, who was an old acquaintance, in Belgium; Giorges Teissier, a marine biologist, who served as director of the Roscoff Marine Station (SBR), in France; and Tyor in West Germany. In the UK, he visited the Royal Society of London, the Linnean Society, the Natural History Museum of London, the Zoological Society of London, and the Royal Botanical Gardens, Kew. In the Natural History Museum, he made a research on hydrozoans in cooperation with Harvey, the head of the Biological Department, and Cornelius, the chief of the Cnidaria Laboratory. At any rate, we have no choice but to feel admiration for the energetic spirit of study that he exhibited. The Royal Society of London and the Natural History Museum of London nominated him as a fellow and an honorary fellow, respectively.

During the period from September 30th to October 14th, 1975, Emperor Showa and Empress Kojun visited the US in return for President Ford's visit to Japan the previous year. It was just after the Vietnam War had ended, and President Nixon had resigned in the wake of the Watergate scandal. Prior to the visit to the US by the Emperor and the Empress, the citizens of the US seemed to have the attitude that they were merely another couple of very important persons, who visited one country after another every year, and therefore paid less attention to them. In contrast to their visit to Europe, however, they were welcomed with no noticeable opposition movement. It seems that the citizens of the US, who have a comparatively short history as a nation, tend to have romantic notions about titled people such as Royal and noble families.

During this visit to the US, the Emperor gave an address at a welcome banquet held at the White House on October 2, using the wording "that most unfortunate war, which I deeply deplore" and expressing his gratitude to the people of the United States for the friendly hand of goodwill and assistance given to Japan for post-war reconstruction immediately following the war "I believe your tolerance and goodwill will be passed on permanently from generation to generation in Japan." Partially thanks to a suitable (or at least appropriate) interpretation by Hideki Masaki, a general official of the Imperial Household (the eldest son of Army General Jinzaburo Masaki), his speech made a deep impression on the US citizens.

During their long stay, the US citizens' interest in the Imperial couple had grown so steadily that the famous New York Times featured photos of them on its front page for six days. Seeing the austere and sincere character of Emperor Showa, who is seen as the symbol of the unity of the people of Japan, with their own eyes, the Americans, unlike the people of other countries, cleared the air between both countries during and after the war, demonstrating that the visit to the US by the Imperial couple had achieved great success. None of prime ministers of Japan who had visited the US earlier had produced such a fruitful result.

By the way, what kinds of reaction did American biologists give? According to private letters from Patricia Morse, who works at Friday Harbor Laboratories and is very well acquainted with Japanese biologists, and from her friends, their reaction to the Emperor's visit may be summarized as follows:

Ahead of Emperor Showa's visit to the US, David Pawson, a chief researcher at the Division of Invertebrates, National Museum of Natural History, Smithsonian Institution, Washington, held several meetings that included the staff members from the Japanese Embassy at the office of Richard S. Cowan, the then-general director of the museum. It was ultimately decided that he would meet the Emperor with David Bayer, an authority on deep-sea octocoralliaprecious corals (*Corallium* spp.) and guide him to the Division of Invertebrates. Hidemi Sato, a graduate of the Department of Zoology, Kyoto University, who was doing research on cell division at the University of Pennsylvania, was appointed as an interpreter, owing to his specialized knowledge of biology. Incidentally, Sato later served as the director of the Sugashima Marine Biological Laboratory, Nagoya University. Nevertheless, Pawson, unfortunately, was unable to return to the museum in time for the Emperor's visit, because a thick fog delayed his flight coming back from Yugoslavia, where he had been attending a conference on cnidarians.

Joseph Rosewater, an expert on molluscs, was quickly sent in to fulfill Pawson's duty. After attending the reception held when he arrived at Andrews Air Force Base on the morning of December 2nd, the Emperor visited the museum in the afternoon. The Emperor took his own specimens of hydrozoan with him to compare with those exhibited in the Smithsonian National Museum and he was very pleased about the similarities between both groups of specimens. Bayer presented the specimen of Adanson-Okinaebisu (*Entemnotrochus adansonianus*) to the Emperor. (Later, Bayer came to Japan to visit to the Biological Institute of the Imperial Household and presented the Emperor with a specimen of one kind of hydrozoan collected from the deep sea in Panama Bay. The Emperor recorded this hydrozoan as a new species and named it *Hydractinia bayeri* to honor his achievement.) The Emperor, on a visit that was supposed to last about 20 min, ended up staying there for over one hour. That same night, the Emperor attended the welcome dinner party at the White House mentioned earlier. It is likely that the visit to the museum was one of the very few chances the Emperor had to genuinely relax during his trip.

The Emperor then moved on from Washington, DC, to the Marine Biological Laboratory, Woods Hole, situated near Boston, on October 4th. This is a laboratory that enjoys tremendous admiration among researchers all over the world who specialize in biology, especially marine organisms, and, during the summer holidays, researchers come there from all corners of the nation. I myself have visited there to conduct research and attend research meetings a number of times. Jim Ebert, a developmental biologist, who was the then-director of the laboratory, welcomed the Emperor and asked him to sign a guest book that had been ceremoniously placed in advance. After the Emperor had signed the book, Ebert asked him if he would like some refreshment, but the Emperor said that he wanted to observe the researches being conducted in the laboratory as soon as possible and, accordingly, the staff quickly guided him to the exhibition hall, where a number of microscopes had been placed. This was, no doubt, a very precious experience for him, seeing as he took the time off from his busy schedule of official duties to do it (Fig. 1.19).

In this room, Sears Crowell showed the Emperor both living hydrozoans and the fixed specimens of hydrozoans, which had been collected from the deep sea, and the Emperor expressed a strong interest in them. Accordingly, Hidemi Sato was

Fig. 1.19 Emperor Showa observing specimens at the Marine Biological Laboratory, Woods Hole, Massachusetts, US, along with his interpreter, Hidemi Sato (on his left side), in 1975 (Courtesy of the National Museum of Nature and Science)

terribly busy with his duty as translator. There is another amazing story that is worthy of special mention. At that time, Shinya Inouè, who is now over 90 and still continues to make his research in this laboratory, showed a film of a series of images of cell divisions being observed under a microscope. Inouè, who was a pupil of Katsuma Dan (later the president of Tokyo Metropolitan University, and also the second son of Baron Takuma Dan, who was shot to death by an assassin in the May 15 incident), developed his own original polarizing microscope. He succeeded in using his microscope to observe the process in which chromosomes are attracted towards opposite poles by spindle microtubules in a living cell at the stage of mitosis. This was a world premiere accomplishment in this field. Inouè was later awarded the International Prize for Biology for achieving a series of researches on this subject. According to some sources, all those present were thrilled by the scenes of his film, the Emperor in particular. One of the viewers of the film was Patricia Morse, an expert on sea slugs (Opisthobranchia). She told me that Anthony Morse, an ancestor 10 generations before her, was an ancestor 8 generations before Edward Sylvester Morse, who served as the first professor of the Department of Zoology, University of Tokyo. She visited Japan several times and gave high praise to "Opisthobranchia of Sagami Bay," which is an illustrated book of the Emperor's collection of organisms, edited by Kikutaro Baba, another expert in the same field.

The Emperor and the Empress stayed in New York for three days starting on October 4th, visiting the New York Botanical Garden on October 6th. The Emperor, who was, at the time, in the process of writing "Flora Suzakiensis," which focused on the plants growing around the Suzaki Villa at Shimoda, observed, with deep interest, some of the plant specimens collected from all over Japan, from Okinawa to Hakodate, which had been brought to the US on the "Kurofune" from Shimoda, led by Perry. At that time, William Steere, an authority on Bryophyta and an honorary curator of the garden, and Tetsuo Koyama, who worked at the garden and later became the director of Makino Botanical Garden, Kochi, gave lectures in the presence of the Emperor. During the lectures, the Emperor and the lecturers kept up a lively conversation, almost as if they had prepared to do so in advance. While the Emperor attended the lectures, Empress Kojun visited a class off-campus who were learning about useful plants (quoted from the book by Koyama [36]).

After their stay on the east coast, the Royal couple went to the west coast, with the Emperor visiting the Scripps Research Institute, a world-famous institute situated in La Jolla, San Diego, on October 9th. It is said that the story about "Biologist Hirohito," regarding the Emperor's stay at this place, has been passed on from generation to generation. One scientist who was present there said:

> One can recognize whether a person is a genuine researcher or not simply by seeing the way in which they hold a specimen and observe an exhibit. His Majesty Emperor Hirohito is truly a genuine researcher! ("A Family of Noble Biologists" [30]).

The Scripps Research Institute presented *Neopilina galatheae*, a species related to chitons (Polyplacophora), which is well known as a living fossil, to the Emperor. He also visited the San Diego Zoo and the San Francisco Botanical Garden. In addition, on October 11th, he visited Bishop Museum, which exhibits a collection focused on the nature and cultures in the Pacific-Ocean region around Hawaii, while stopping over in Hawaii on the way back from the visit to the US. According to Ryuzo Yanagimachi, a professor at the University of Hawaii, who himself has won the International Prize for Biology, the specimens of hydrozoans about which the Emperor had inquired had previously been sent to the mainland of the US; the staff of the museum hastened to request that the other party return the specimens, which thankfully happened in time, and they were thus able to show the returned specimens to the Emperor.

The stories mentioned above may elicit a misunderstanding that the Emperor mainly spent his time visiting museums and institutes specializing in biology, but that is not correct at all. The fact is that, as was the case with his visit to Europe, he made full use of the short, precious time afforded him each day, some 20 min to one hour, carved out of his terribly busy official duties, to contribute to academic exchanges. According to sources close to the Emperor at the Imperial Household Agency, who were responsible for scheduling the Emperor's official duties, the Emperor told:

> I intend to express my gratitude to the US Government for continuing to extend its helping hand to Japan in its time of great difficulty as we have recovered, and to so many citizens of the US for showing understanding of such a governmental policy during my visit to the US. I do not in any way intend merely to enjoy my hobby (research on biology) and visiting Japanese-Americans (Miwa [42]).

When they had a chance to come into close contact with the Emperor, the citizens of the US, who, as mentioned earlier, had shown almost no interest in the Emperor's visit up to that point, were impressed by his stories of a number of experiences, including those involving heartache, recognizing him as a "living record of history," and gained a more favorable viewpoint of him as a result of having gotten a glimpse into his sincere personality as a dedicated "marine biologist." They saw this episode as reflecting the serious and honest personality of the Emperor.

Emperor Showa's Later Years

As mentioned above, after accomplishing his heartfelt wish of visiting the US and Europe, the Emperor resumed his earnestly dedicated work on his research into hydrozoans and plants. Many scholars continued to deliver lectures to the Emperor in the interest of his strong desire to learn. It is said that a Scientific Committee and a Cultural Committee, both of which consisted of specialized scholars, had been formed after World War II, and the competent members of each committee had been chosen to deliver lectures. The members of the Cultural Committee delivered lectures on world affairs and the historical background, while those of the Scientific Committee, including Yo Okada, spoke mainly on biology. When the Emperor made an Imperial tour throughout the country, new books related to this tour, edited by the Biological Laboratory of the Imperial Household, were issued, and the Emperor always took the time to discuss the material to be featured in these books with the persons who lived in the places he visited and the authors of the books. In the Emperor's later years, the lectures by the members of these committees were re-organized into "Integrated biological lectures," and several members who were animal or plant biologists were always invited to deliver their lectures on the most current knowledge, especially taxonomical knowledge, to the Emperor. These "Integrated biological lectures" were continued up until the year right before the Emperor passed away.

Soon after turning 70 years old, the Emperor began to display various kinds of symptoms of aging, and the chamberlains, including Irie, made arrangements to reduce the burden on him in regard to official duties as much as they could. Even later in life, the Emperor sometimes appeared to walk with unsteady steps. However, Hiroshi Ito, one of the members of a botanist group that worked with the Emperor, who visited the Nasu Imperial Villa in the year before he passed away, said;

> We are concerned about the Emperor's walking with unsteady steps when we watch him on TV, but he recently walked sprightly along a mountain path while conducting a survey of plants (quoted from "A Family of Noble Biologists" [30]).

It is possible that the reason why the Emperor appeared to walk with unsteady steps on TV is that he always took great care not to slip while walking down steps. This holds true for myself as well, having almost reached the same age as the Emperor at that time, when I go up and down stairs. And yet, curiously, I still get energized immediately when I am collecting butterflies, my own hobby.

In 1985, various kinds of events celebrating Emperor Showa's 60-year reign were conducted. That year, the Emperor turned 84 years old, older than the 108th Emperor Go-Mizunoo, who had held the record for living the longest by then. In the same year, grand chamberlain Sukemasa Irie, who continued to introduce a vast variety of information on the Imperial Family that had so far been veiled to the public, decided to quit his post, taking the opportunity upon turning 80 years old and getting the approval of the Emperor. Sadly, he passed away suddenly only two days before the formal order of retirement was issued. The Viscount Irie family are the descendants of Teika Fujiwara (a famous waka poet), and Tamemori Irie, Sukemasa's father, served Emperor Showa in his Crown Prince days as the Crown Prince's grand chamberlain. Tametoshi, the son of Sukemasa, was a classmate of the present Emperor, Akihito. In addition, Sukemasa was appointed as a chamberlain while working at Gakushuin University as a professor. Having graduated from the Faculty of Literature, University of Tokyo, he was known as a good writer and prominent essayist. The officials who represent the Imperial Household Agency, including the grand steward of the Imperial Household Agency, who are usually appointed from among governmental officials, are often loathe to disclose information on the Imperial Family to the public at their discretion, due to their positions. In contrast, Sukemasa had disclosed an exceptional amount of information on the Imperial Family to the public, taking advantage of his career and position and contributing to the reduced distance between the Imperial Family and the public.

Through the processes described later, in 1985, the "International Prize for Biology" was established to encourage advances in biology in commemoration of Emperor Showa's 60-year reign and long-term devotion to his research on biology. The first International Prize for Biology was granted to Eldred John Henry Corner, an English botanical taxonomist, who was well acquainted with Japanese researchers. Crown Prince Akihito, on behalf of Emperor Showa and Princess Michiko, attended the award presentation ceremony (at present, Their Majesties the Emperor and Empress attend the ceremony). Later, Corner visited the Imperial Palace and had an audience with Emperor Showa. Since then, the International Prize for Biology has been granted to the leading biologists throughout the world who have made the highest-ranked achievement in the fields of taxonomy and systematic biology, in which the Emperor specialized, and other researches. The prize marked its 30-year anniversary as of 2014. In the same year, the Japan Prize, which is granted to researchers who have made a superior achievement in the science and technology field, was also established. The Emperor and Empress have attended the award presentation ceremony ever since the first one.

At his birthday party in 1987, the Emperor got physically ill, and, during his stay at Nasu in the summer, he was still not feeling well, resulting in his being admitted to the hospital for surgical treatment in September. It was historically the first time in Japan that the Emperor underwent a surgical operation with a scalpel. The Emperor's pathological condition was announced as chronic pancreatitis, not cancer. According to the diary written by Ryogo Urabe, a chamberlain who had served the Emperor since 1969, the Emperor eagerly desired to resume both attendance at constitutional functions and his research on biology after the surgery. To respond to his desire, the microscopes and tools for observation were moved to the Fukiage Imperial Palace from the Biological Institute of the Imperial Household to prepare for his resumption of his research in January 1988. On behalf of the Emperor, the Crown Prince and Princess attended the "1988 National Tree-Planting Ceremony." In the same year, the Emperor visited the Suzaki Imperial Villa twice and, in the summer, stayed at the Nasu Imperial Villa for a long period to conduct a survey of plants. On August 13th, he returned to Tokyo by helicopter, and on the 15th, he attended the "Memorial Ceremony for the War Dead" to make a speech. At the end of August, a meeting with the authors of "Flora sedis Imperatoris Japoniae" was held at Nasu, which turned out to be the last opportunity that the Emperor had for interaction with biologists. The Emperor wrote a waka poem at Nasu in the same year:

At the break of day, a woodpecker's pecking sound is heard, then fades as the silence comes.

On September 12th of the same year, the Emperor visited the Biological Laboratory of the Imperial Household, in what would turn out to be his last visit there. One week later, he vomited a large amount of blood and was given an urgent blood transfusion. Even under such a severe pathological condition, the Emperor was still concerned, on his sickbed, about the distributions of "The Hydroids of Sagami Bay" (author: Hirohito) that had just been issued and the work on proof-reading the book "Flora Sedis Imperatoris Japaniae," to be issued soon. For example, the Emperor gave detailed instructions as to what was still to be done, such as confirmation of the consistency in the scientific name of the down redwood (*Metasequoia glyptostroboides*) between the preface and the text body. It was recorded by chamberlain Urabe, who was present for both the demise of the Emperor Showa and the New Emperor, Akihito's, accession to the throne, that this process had continued until the beginning of November [40]. I take my hat off to the Emperor for his unimaginable dedication as a researcher. Emperor Showa passed away at the age of 87 in the early morning of January 7th, 1989.

Achievements of Emperor Showa

As mentioned above, Emperor Showa had mainly collected marine invertebrates from Sagami Bay and studied them for many years, but he entrusted the publication of the results of his research on collected organisms, except for hydrozoans, to the experts who specialized in the field of marine biology. These reports on his research, edited by the Biological Institute of the Imperial Household, were published one after another (Fig. 1.20). Among these books, "Opisthobranchia of Sagami Bay," with explanations by Kikutaro Baba of Osaka Gakugei University (the current Osaka Kyoiku University), was first published by Iwanami Shoten in 1949. Recently well known to the public as a result of having been shown on TV, the brilliant colors of the beautiful opisthobranchs belonging to the phylum Mollusca, called sea slugs, are reproduced vividly in the illustrated book. My friends and I, having just entered the university, were attracted to the beauty of the organisms that appeared in this book when we referred to it during our practical study of the art of collecting samples along the beach at the Misaki Marine Biological Station. The books that were subsequently published are listed chronologically below.

"Ascidians of Sagami Bay" (Takashi Tokioka, a professor of Kyoto University), 1953, Iwanami Shoten
"Opisthobranchia of Sagami Bay (Supplement)" (Kikutaro Baba), 1955, Iwanami Shoten
"The Crabs of Sagami Bay" (Tsune Sakai, a professor at Yokohama National University), 1965, Maruzen
"The Hydrocorals and Scleractinian Corals of Sagami Bay" (Motoki Eguchi, a professor at Tohoku University)
"The Sea Shells of Sagami Bay" (Tokubei Kuroda, Tadashige Habe, Katsura Oyama, all doctors of science), 1971, Maruzen
"The Sea-stars of Sagami Bay" (Ryoji Hayashi, a professor at University of Toyama), 1973, Hoikusya
"The Crustacean Anomura of Sagami Bay" (Sadayoshi Miyake, a professor at Kyushu Sangyo University), 1978, Hoikusya

Fig. 1.20 Books published by the Biological Institute of the Imperial Household and the papers written by Emperor Showa (Courtesy of the Imperial Household Agency)

"On the Genus *Ephelota* (Ciliophora, Suctoria) from the Coasts of the Izu Peninsula and Niijima" (Ryozo Yagyu, a professor emeritus at Hiroshima University), 1980, Hoikusya

"The Brittle-stars of Sagami Bay" (Seiichi Irimura, a teacher at Yokohama City Totsuka High School), 1982, Maruzen
"The Sea Urchins of Sagami Bay" (Michio Shigei, a professor at the Kyoto Institute of Technology), 1984, Maruzen
"The Sea Spiders of Sagami Bay" (Koichiro Nakamura, a postgraduate student, the Marine Research Laboratory, University of Tokyo), 1987, Maruzen
"The Demospongiae of Sagami Bay" (Senji Tanida, an emeritus professor at Hokkaido University), 1989, Maruzen

The names in parentheses indicate the professionals who compiled the individual books. Incidentally, ascidians (sea squirts) belong to the class Ascidiaces, sea stars belong to the class Asterozoa, crustacean anomura (hermit crab) belong to Paguroidea, Suctoria is a special Ciliophora (ciliates), brittle stars belong to the class Ophiuroidea, sea urchins belong to the class Echinoidea, and sea spiders attached to seaweed belong to the class Crustacea. Just for the readers' information, "Opistobranchia of Sagami Bay" was revised and reissued with supplements (Kikutaro Baba, 1990, Iwanami Shoten). Besides the series of 12 books listed above, "Crinoid Fauna (Echinodermata) of Sagami Bay" (Ichizo Kogo and Toshihiko Fujita, 2014), focusing on the specimens transferred to the National Museum of Nature and Science from the Biological Institute of the Imperial Household, was published by Tokai University Press.

Moreover, the following books on plants of which Emperor Showa was co-author, except for those on myxomycetes, were also published:

"Flora Nasuensis," 1962, Sanseido
"Flora Nasuensis Additions & Emendations," 1963, Sanseido
"Additional Notes to the Myxomycetes of Nasu" (Hirotaro Hattori), 1964, Sanseido
"Nova Flora Nasuensis," 1972, Hoikusya
"Flora Suzakiensis," 1980, Hoikusya
"Nova Flora Nasuensis Supplementum," 1985, Hoikusya
"Flora Sedis Imperatoris Japoniae," 1989, Hoikusya

As described above, the Emperor had his first experience proofreading a book from the beginning with "Flora Nasuensis," one of the books on plants, and whenever the publication of one of these books was planned, the Emperor would have discussions with the five botanists and others from the planning stage on, either in the Imperial Palace or at the Imperial Villas. As already described, the last book was not published while the Emperor Showa was still alive. The co-authors' names of the books are as follows: for "Flora Nasuensis" and "Nova Flora Nasuensis," Tatsuo Sato and the five botanists; for "Flora Suzakiensis," Bungo Kawamura (instead of Tatsuo Sato) and the five botanists; for "Nova Flora Nasuensis Supplementum," the five botanists, excluding Masaji Honda; and for "Flora Sedis Imperatoris Japoniae," Arika Kimura, Shiro Kitamura, Hiroshi Ito, and Bungo Kawamura, excluding Hiroshi Hara.

Sample collection, sample or specimen observation and the acquisition of relevant literature had been a continuous part of Emperor Showa's life for many years.

Nevertheless, he had never published his papers under his own name, the Emperor Hirohito, until the age of 65, in February 1967, three years after the Tokyo Olympic Games were held, when he published his first paper at Hirotaro Hattori's suggestion. At that time, everyone was usually forced to retire when reaching the age of 65, no matter where he or she worked, be it a public or private organization.

The papers of the Emperor published in those days were limited to hydrozoans. The first paper, entitled "A review of the hydroids of the family Clathrozonidae with descriptions of a new genus and species from Japan," was published by the Biological Institute of the Imperial Household. In this paper, he corrected the existing taxonomic classification of Eberhard Stechow, who was an authority on the taxonomy of hydrozoans, and added a new species, *Pseudoclathrozoon cryptolarioides*, in a new Genus, *Pseudoclathrozoon*, to the taxonomic description. At the suggestion of Hatsuki Tsujimura, who was in charge of the schematic pictures in the paper, part of a colony of *Pseudoclathrozoon cryptolarioides* was designed by Sagenji Yoshida, a professor emeritus at Tokyo University of the Arts, to be featured on the medal awarded to the winners of the International Prize for Biology.

The Emperor often seemed to have doubts about Stechow's taxonomic system of classification and, as early as 1948, during a discussion with scientists, including Taku Komai of Kyoto University, he told of his great efforts to solve this difficulty:

My classification is different from Stechow's sometimes. The differences have not been clarified, even after repeating a closely comparative examination between both classifications.

Reflecting upon the Emperor's doubt, Taku Komai said:

I think the Emperor's classification is correct. Although Dr. Stechow is a respected biologist, who has exerted his utmost efforts, unfortunately, he doesn't seem to have been able to collect as many samples as the Emperor has. My Emperor, you have carried out your research taking full advantage of your abundant collection of valuable materials, so I believe it would be best for you to continue to conduct your research in your own way. Out of all the countries in the world, the widest variety of organisms live in Japan, and Sagami Bay, especially, is a treasury of organisms, a so-called Natural Museum of Organisms. It is natural that your research on the specimens collected from the bay should be highly esteemed (Tanaka [60]).

Besides Hattori, both Komai and Toru Uchida, an authority on jellyfish, who was a professor at Hokkaido University, prevailed on the Emperor to publish his own paper, telling him:

It is usual practice in academic societies that, when an error is recognized, it should be corrected. Accordingly, it is possible that a hypothesis that you have developed will be corrected by another researcher, who will formulate his new hypothesis in the future, if any comes to be. It is an ideal situation for us to further advance our researches through the friendly-rival relationship between researchers.

Following their suggestions, the Emperor finally made up his mind to publish the paper under his name, Hirohito (quoted from the book edited by Kagaku Asahi [30]).

Furthermore, the following eight papers listed below were also published under his name, both in Japanese and English:

"Some hydroids of the Amakusa Islands," 1969
"Additional notes on *Clathrozoon wilsoni* Spencer," 1971
"Some hydrozoans of the Bonin Islands," 1974
"Five hydroid species from the Gulf of Aqaba, Red Sea," 1977
"Hydroids from Izu Ôshima and Niijima," 1983
"A new hydroid *Hydractinia bayeri* n.sp. (Family Hydractiniidae) from the Bay of Panama," 1984
"The hydroids of Sagami Bay," 1988
"The hydroids of Sagami Bay. II. Thecata" (supplemented and revised by Mayumi Yamada), 1995

The Emperor's papers on other species of organism were limited to those from Sagami Bay, but the papers on the hydrozoans in which he specialized reported on those from a wide range of sea regions, from Amakusa and Ogasawara in Japan to overseas, including specimens that had been brought as offerings to the Emperor.

The last two listed monographs, into which Emperor Showa's researches on hydrozoans were compiled, are voluminous editions, numbering roughly 300 and 600 pages (Japanese and English versions, respectively). The former monograph dealing with Athecata was published immediately prior to the demise of the Emperor Showa. In the latter monograph on Thecata, published after his demise, Mayumi Yamada, who made great efforts in realizing its publication, compiled almost all of the Emperor's posthumous manuscripts, including his original drawings, with almost no modification. The former monograph describes 67 species of hydrozoan, consisting of 17 new species and 10 newly-recorded species from Japan; and the latter covers 184 species, consisting of 10 new species and 19 newly-recorded species from Japan. Looking at either of these two monographs, one can see that the Emperor's papers are filled with elaborate and detailed descriptions, reflecting his character. It is quite certain that, in the future, these will continue to be fully used as important pieces of literature by researchers all over the world working on the same kinds of subjects.

It should be noted that the English versions of the Emperor's papers were proofread by Jean Clark Dan (Japanese name, Jinko Dan), a professor at Ochanomizu University for many years. Jean Dan became acquainted with Katsuma Dan, who later became a world authority on the mechanism of cell division, at the Graduate School of the University of Pennsylvania, married him, and accompanied him back to Japan. During World War II, she brought up five children in Japan and, after the war, she discovered an important phenomenon at the Misaki Marine Biological Station, that is, the sperm acrosome reaction essential to the fertilization of animals. In her later years, she acquired Japanese citizenship, worked as a professor at Ochanomizu University, and then served as the director of the Tateyama Marine Biological Laboratory. She proofread the English versions of papers of a great number of biologists in Japan besides the Emperor's, including my own, and, moreover, she also proofread a couple of academic books, entitled "Development of Invertebrates, Volumes 1 and 2" and "Genbaku no Kora" ("Children of Hiroshima") (unfortunately, the latter could not be published in the US), translated into English from Japanese, continuing to make a great contribution as a bridge between Japan and the US, as well as between Japan and Europe, until she passed away in 1978.

Evaluation of the Achievements of Emperor Showa

Emperor Showa was a humble man, saying, "I conduct my biological research in spare moments from my official duties, so it is only a hobby" (quoted from the book "Observers of the Imperial Court 'School'" by Irie [24]). However, it can be clearly understood from the descriptions above that this was not so. The primary subject of his research, on which the Emperor focused as his specialized field, was hydrozoans. The Emperor himself wrote papers on hydrozoans. His secondary subject was myxomycetes. The Emperor preferred, as a hobby, researches on seed ferns (Pteridophytes), seed plants (Spermatophytes), seaweeds, and a great number of marine organisms. However, despite the Emperor's modesty, his significant knowledge of these plants and organisms was clearly at the professional level, not the amateur level, as seen in the fact that he carried out such thorough investigations of them. As described earlier, he also observed seasonal and annual changes in plants.

According to Sukemasa Irie, the Emperor made it his lifes' work to conduct taxonomic research on hydrozoans, realizing that his research on plants was at the non-professional level [26]. However, typical floras tend to cover only the records of plants, while the Emperor's flora accurately describes the particular characteristics of plants, for example, when their flowers have come into bloom and when they have fallen, on the basis of his actual observations in the same locations and at the same times every year. The Emperor's research on plants is thus to be far more highly esteemed in terms of quality than those of botanists who spend just a few days walking around observation points (quoted from the book by Tei [61]).

Giving some examples of overseas evaluations, in 1951, shortly after "Opisthobranchia of Sagami Bay" was published, George Evelyn Hutchinson of Yale University enthusiastically praised the monograph, and wrote, in "American Naturalist":

> The most sumptuous zoological monograph ever to arrive at the office of the *American Naturalist* is devoted to them (opisthobranch molusks), and is the first of a series of reports on material collected in Sagami Bay by His Majesty the Emperor of Japan.

He also mentioned that, among the organisms recorded in the monograph, amounting to 155 species, the Emperor himself collected almost all save for 11 of them, and concluded:

It is greatly to be hoped that subsequent volumes from the Biological Laboratory of the Imperial Household will not be long in appearing [19].

In the Message to "The Emperor's Biological Research" published by the National Museum of Nature and Science for the Special Exhibition celebrating the 110th anniversary of the museum, Frederic M. Bayer, an authority on Cnidaria at the Smithsonian Institution stated that:

His Imperial Majesty Hirohito, Emperor of Japan, is therefore unique among world monarchs. Not only does he personally collect specimens from all kinds of habitats, from mountains to seashore and even the depth of the sea, but, in addition to inviting the collaboration of foremost contemporary specialists in the tradition of earlier monarchs elsewhere in the world, he also studies them himself. It would be neither fair nor accurate to describe Hirohito the Emperor as a scientific dilettante who devotes only a small part of his leisure to his scholarly studies.

He then added:

But most important, the Emperor has personally conducted investigations of his favorite marine animals, the Hydroida. From a scientific standpoint, these are among the more difficult animals to study and classify, because many of them have elaborate life cycles that include two vastly dissimilar generations, each of which has been named and classified independently. It is a tribute to his scientific determination and skill that he has pursued the difficult study of these perplexing organisms with exceptional success, and has published the results under his own name [5].

The "two vastly dissimilar generations" written of herein indicate those of jellyfish and polyps, as described earlier. In addition, Edred Corner of the University of Cambridge stated, in his memorial address commemorating the deceased member of the Royal Society [9]:

……It was far more. No amateur could have encompassed and mastered the vast field of nature that he did and have risen to a world authority. His enjoyment of biology… not only provided comfort and relaxation, as others have remarked, but reflected his confidence in natural science as just a means for uniting all mankind. With much resources than others, he assembled a biological court of advisors, in whom he had implicit trust, and became the first emperor to have devoted his spare time to science.

Moreover, it is said that, at the Institute of Marine Biology, Russian Academy of Science in Vladiostok, Emperor Showa's papers and a series of Sagami Bay monographs on hydrozoans have been similarly fully used as essential literature for the researches at the institute.

As known from the descriptions seen so far, Emperor Showa's contribution to biology has been far more highly esteemed in other countries than in Japan. Needless to say, some have conflated it with accusations regarding the Emperor's responsibility for World War II, but this issue must be discussed separately from his achievements in the field of biology. The Emperor himself seemed to give priority to official duties and rituals, and was mindful of keeping his research on biology under wraps. This may also be applicable to the current Emperor and Prince Akishino. In the summer of 1988, the year before the demise of the Emperor Showa, a special exhibition entitled "His Majesty the Emperor's Biological

Researches" celebrating the 110th anniversary of the National Museum of Nature and Science was held. About 300 exhibits, including the specimens collected by the Emperor, his books and microscopes, and many photographs were put on display for more than a month. However, fewer than 60,000 people visited the exhibition, even though it was held during the summer vacation. The number of visitors was far less than that for the "Dinosaurs Exhibition," which enjoyed tremendous success whenever it was held. Such an unsatisfactory result is regrettable, but not particularly surprising, in part due to Japanese citizens' lack of interest in taxonomy and biology in general.

One of the features of the Emperor's research was to limit the location of collection and observation of marine organisms to Sagami Bay, except for his beloved hydrozoans (including specimens offered by other countries and those from Ogasawara) and myxomycetes, while the study of plants was confined to the Imperial Palace, Nasu and Suzaki. Excluding the special case when he used warships before World War II, even after a large-sized collection boat had become available, he never disobeyed this rule. It was fortunate for the Emperor that Sagami Bay is one of the world's great treasuries of marine organisms, and he made all possible efforts to acquire information on the animals living and the plants growing there. Accordingly, the monographs compiled by the experts in their respective fields are valuable materials for the researchers of the world. The observatory works on plants at Nasu and Suzaki came to fruition, resulting in a lengthy record of vegetation in these limited locations. Needless to say, during that period, the Emperor also discovered a number of new species within an extremely vast range of animals and plants.

Among his collection, in terms of hydrozoans, 32 new species in total, 1 new variety and 61 newly-recorded species from Japan are contained in his books, and 19 new species and 1 new variety were recorded by researchers in other countries. In terms of other organisms, he also discovered the following new species: Myxomycetes: 5 new species and 3 new varieties; naked sea slugs (Nudibranch): 67 new species and 1 new subspecies; crabs (Brachyura): 24 new species and 1 new subspecies; starfish (Asteroidea): 7 new species and 1 new subspecies; shellfish: 110 new species or subspecies; and seaweeds: 24 new species (quoted from the private letter of Hiroshi Namikawa). Besides Linné, who was the first to propose a binomial nomenclature, some of the biologists who competed with each other in discovering new species of animals and plants during the Age of Discovery, and some taxonomists and those who have worked on the classification of certain groups of organisms, like Junichi Aoki, described earlier, have recorded a far greater number of discoveries than the Emperor. In the 20th century, however, probably none have contributed as much as he to the discovery of new species, particularly, it should be remembered, in as much as he racked up his achievements within a limited sphere of activity due to his position. Hatsuki Tsujimura, who served the Emperor and was deeply involved in his research for many years, has said that the Emperor's worth as a biologist can be clearly seen in the fact that he once said, "I am not one of those specialists whose knowledge is restricted to that of his or her own research subject" (quoted from the book by Sukemasa Irie, described earlier [24]).

In the issue of *Asahi Shimbun* from January 8th, 1989 (the first year of Heisei), Mayumi Yamada summarized the situation as follows:

> The Emperor conducted an exhaustive research based on observation of the specimens collected within a limited sea area, Sagami Bay, over a long period. Almost no researcher in any other country matches the Emperor in their elaboration of research. Accordingly, the Emperor's research may be of great academic significance. The Emperor referred to a wide range of literature from around the world and took a cautious and profound stance in regard to his research. Needless to say, his collection of a wide variety of specimens stored at the Biological Institute of the Imperial Household is perhaps the largest in Japan, and I believe that it is of great international value. The rich variety of organism specimens collected in Sagami Bay during his research on hydrozoans that he has provided to biologists specializing in the field also represents an irreplaceable contribution to the academic society.

Empress Kojun and Former Imperial Princesses and the Study of Biology

Married couples who are engaged in the same field, such as the Curies, can be found here and there in history. These couples are able to conduct their researches smoothly, because, sharing common knowledge in the same field, they can easily understand what their partners are thinking. Even in cases in which the partner is not a researcher her/himself, they can still be an encouraging supporter of the researcher as long as he or she is interested in the researcher's subject. Empress Kojun served the Emperor not only as a partner in daily life and official duties, but also as a reliable supporter, who was no less interested in the living organisms and seaweeds growing along the beaches, weeds and flowers, and mosses and mushrooms than the Emperor. Empress Kojun was regarded as a leader of the "Observers of the Imperial Court 'School,'" consisting of the chamberlains, including the former grand chamberlain Sukemasa Irie.

The tennis court at Karuizawa is well known for providing the present Emperor, Akihito, and Empress, Michiko, with an opportunity to meet with and come to understand each other closely. Emperor Showa and Empress Kojun also often enjoyed playing tennis and golf in each other's company in their early days. It is said that Empress Kojun was better at tennis than the Emperor. The Imperial Couple was harmonious and, according to Osanaga Kanroji, who served Emperor Taisho and Emperor Showa for decades, they almost never quarreled with each other (quoted from the book by Kanroji [33]); indeed, evidence suggests that the two got into an argument only once. Besides walking around the Imperial Palace, the Empress went along with the Emperor almost every time he went out to Nasu and Suzaki to collect animals and plants. This practice continued until it became difficult for her to walk, due to a backache in her later years. At Hayama, the Empress went along with the Emperor when he collected samples along the beach, as well as when he used a collection boat, unless he was going to a distant place. The Empress also went along with the Emperor when he went on the boat to the Misaki Marine Biological Station, a short distance from Hayama.

The book "Emperor Showa and His Biological Research," written by Toku Tanaka and published in 1949 [60], features photos of heartwarming scenes, in which the Emperor and the Empress refer to an illustrated reference guide of flora together. During the period from 1935 to 1941, Empress Kojun actually prepared the specimens of seaweeds collected at Hayama herself. These were collectively called "The Empress's herbarium," and were later combined with other specimens. From "The Empress's herbarium," Yukio Yamada of Hokkaido University, who was an expert on seaweeds, recorded a dozen new species. Moreover, the Empress went out of her way to prepare the specimens of moss that the Emperor had not gotten around to, under the guidance of an expert at Nasu. In addition, there are photos of the Empress classifying shellfish at the Fukiage Imperial Palace (Fig. 1.21). As is known from the photos and the descriptions above, the Empress essentially held the position of a co-researcher, although there are no papers with her name as a co-author. The Empress liked growing roses, and pruned her rose bushes herself. Emperor Showa desired to reflect the natural environment of Musashino as much as possible in Fukiage Imperial Palace, but did not complain to the Empress about her rose garden.

The Empress had a variety of hobbies, such as calligraphy and painting, writing waka poems, embroidery, dyeing, noh chanting (recitation), and playing violin and

Fig. 1.21 Empress Kojun sorting out shellfish (Courtesy of the Imperial Household Agency)

piano, along with playing sports, although, unlike many of her peers, she did not engage in horseback riding. The Prefatory Note to "Flora nasuensis" features a Japanese-style painting of an Asian blue lobelia (*Lobelia sessilfolia*) made by the Empress, which was printed in a newspaper at that time. Before getting married, the Empress learned the art of Yamato-e-Painting (a traditional Japanese style of painting) from Wakanari Takatori, who was an expert on painting the figures of historically well-known persons, and afterwards, studied further with Gyokudo Kawai and Seison Maeda, both of whom received the Order of Cultural Merit. She created her works under the Gago (pseudonym) of 'Toen' and issued "Book of Paintings, Toen (Toen Gasgu)" (1967), "Book of Paintings, Kinpo (Kinpo Shu)" (1969) and "Revised and Enlarged Version of Kinpo Shu" (1989). The Empress compiled her paintings, based on the sketches she made of weeds and flowers, often at Nasu, into these books (Kageyama [31]).

Since Empress Dowager Shoken took up the cultivation of silkworms in the Imperial Palace in 1871, this practice has been passed down to the next-generation Empresses, up to the present day. Raw silk and silk fabrics were leading goods in the export industry at that time. However, this event is said to be traceable to the descriptions in "Nihon Shoki" ("Chronicles of Japan"). At present, although almost no silkworm cultivation currently takes place in Japan, it has been regarded as a pair with the rice cultivation taken up by Emperor Showa, and Empresses Kojun and Michiko gained experience in silkworm cultivation at the Momiji-yama Sericultural Station of the Imperial Household. Empress Kojun became the Empress Dowager after the demise of Emperor Showa, and passed away herself at the age of 97 in 2000. She outlived Fujiwara Kanshi (Hiroko), who was the 70th Emperor Go-Reizei's Empress and passed away at the age of 92.

The princesses of Emperor Showa were brought up at their home during their young days, and were not put out to be nursed by foster parents, which was one of the traditional practices of the Imperial Family. However, after suggestions by the royal advisors that "such a non-traditional practice might spoil the princesses," they were individually moved to Kuretake-Ryos (fostering houses) before entering Gakushuin Girl's Junior and Senior High School, so as to be brought up away from home. Princess Teru Shigeko, the eldest daughter of Emperor Showa, was interested in biology and wished to work at the Biological Institute of the Imperial Household, but her desire could not be realized. Following the practice of the Imperial Family at that time, she got married to Prince Higashikuni Morihiro, a member of the Imperial Family and the military personnel. Prince Higashikuni Morihiro, the eldest son of the Prince Higashikuni Naruhiko, who served as the first Prime Minister after the end of World War II, seceded from the Imperial Family and entered the University of Tokyo after the end of World War II. Shigeko Higashikuni continued to be interested in biology after getting married, studying nutria (*Myocastor coypus*) by herself and achieving success in its cultivation, as well as raising the horned worms of hawkmoths. Unfortunately, she died young.

Princess Taka Kazuko, the third daughter of Emperor Showa, got married to Toshimichi Takatsukasa, who was a successor to the Duke Takatsukasa Family, one of five regent houses, and a researcher on railways. Kazuko seemed to be uninterested in biology, but Nobusuke Takatsukasa, Toshimichi's father, was one of the so-called noble biologists and was interested in birds. He entered the Department of Zoology, Faculty of Science, University of Tokyo, and studied under Isao Iijima, the first graduate from the Department of Zoology. In 1912, he set up the Ornithological Society of Japan, together with Nagamichi Kuroda and Seinosuke Uchida, who were his fellow disciples, under the presidency of Iijima, and later succeeded Iijima, becoming the second president. He conducted research on "poultry" and compiled the results of his research into a book. He also purchased the entire series of books "Catalogue of the Birds," written by John Gould during the reign of Queen Victoria, which are not available today. As described above, he contributed to the development of ornithology in Japan. The series was presented to the Yamashina Institute for Ornithology by his wife after he passed away. Later, Princess Nori Sayako conducted a research on Gould's books. Nobusuke also served as the chief priest of the Meiji Jingu Shinto Shrine.

Princess Yori Atsuko, the fourth daughter of Emperor Showa, got married to Takamasa Ikeda, who was born into a family of the feudal lord of Okayama and ran a farm. She also seemed to be initially uninterested in biology. However, Takamasa became the director of the "Ikeda Zoological Garden," which he founded himself later, suggesting that she was perhaps linked to biology by predestined fate. Princess Suga Takako, the fifth daughter of the Emperor Showa, got married to Hisanaga Shimazu, who was the first cousin of Empress Kojun and a fellow student of the present Emperor. He worked at the Export-Import Bank of Japan, then at Sony, and, at present, holds the post of Chairman of the Board of Directors of the Yamashina Institute for Ornithology, indicating that he was also predestined to be involved with living organisms. Moreover, the Imperial Princesses had often joined Emperor Showa and Empress Kojun in walking around and collecting samples at the Imperial Palace and Imperial Villas, sometimes, in later years, even bringing along their own children.

Chapter 2
Emperor Akihito [1933–] and Prince Hitachi [1935–]—The Second Generation Biologists

The Imperial Palace and Villas

Before describing the contributions to biology by the present Emperor, Akihito, and his younger brother, Prince Hitachi, I would like to discuss some historical incentives that have allowed the members of the Imperial Family to become interested in living organisms and biology. Among these incentives, the first is the environment of the Imperial Palace and Villas, where they have resided in the past and still live now. The existing Imperial Palace was called Edo Castle (also called "Chiyoda Castle"), which was the residence of the Tokugawa Shogun in the Edo Period. In the period from 1888 to 1948, it was called Kyujo (Imperial Palace). The name Kyujo is more familiar to me than other names. The area of the Imperial Palace, including the East Gardens, is 1,150,000 m², a little wider than the total area of Tokyo Disneyland and Tokyo DisneySea. In bird's eye aerial photographs of the Imperial Palace, the green space of the Imperial Palace covers a wide range of area in the central part of Tokyo (Fig. 2.1).

The Imperial Palace is formed by the following components: the Residence where the Emperor and the Empress reside; the Palace where various ceremonies and events are held; and the Imperial Household Agency building. Edo Castle was composed of Hon-maru, Ni-no-maru, San-no-maru and Nishi-no-maru within the inner moat, and, on the west side of Edo Castle, there was a garden called "Fukiage," separated from the Castle by the moat Dokan-bori, still the current Fukiage Garden. The eastern area composed of Hon-maru, including the castle tower (Tenshukaku), Ni-no-maru and San-no-maru has been refined into the East Garden. The East Garden houses the Imperial Music Hall, called Tokagakudo, the Museum of the Imperial Collection, called San-no-maru Syozokan, where the arts and crafts inherited by the Imperial Family from generation to generation are housed, and the Archives and Mausolea Department and the Music Department of the Imperial Household Agency. The East Garden is open to the public. When visiting the Palace and Residence, we go through the Sakashita-mon gate, facing

© Springer Nature Singapore Pte Ltd. 2019
H. Mohri, *Imperial Biologists*, Springer Biographies,
https://doi.org/10.1007/978-981-13-6756-4_2

Fig. 2.1 A bird's-eye view of the Imperial Palace. Copyright © National Land Image Information (Color Aerial Photographs) Ministry of Land, Infrastructure, Transport and Tourism (partly modified)

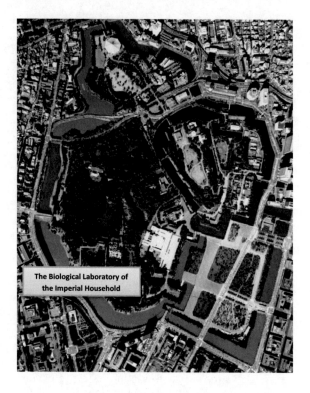

The Biological Laboratory of the Imperial Household

Tokyo Station. The old Nishi-no-maru is composed of the Imperial Household Agency building and the Palace. The Residence is situated in the Fukiage Garden.

The site of the Fukiage Garden is most likely situated over the remains of the residences of Tokugawa gosanke (three privileged branches of the Tokugawa family), but after a great fire occurred in 1659, it was reconstructed into a garden to serve as a vacant lot so as to prevent future fires from spreading. Emperor Meiji first used the Nisi-no-maru Palace as his Residence, but it too was destroyed by a fire. For this reason, the so-called Meiji Palace, integrating the palace and the residence together, was constructed in Nishi-no-maru in 1888. The residence was used by both Emperors Taisho and Showa, but, once again, fire due to flying sparks during an air raid attack in 1945, just before the end of World War II, destroyed the whole thing. During the war, the single-storied "Gobunko," which means "library," was built in the Fukiage Garden, with two basement levels to be used as shelters to protect Emperor Showa and Empress Kojun from air raids. As the tactical situation grew more serious, a stronger annexed structure was separately constructed underground, connected to "Gobunko" by an underpass. The Imperial Council that led to the decision to end the war was held in the annex.

Sympathizing with the citizens ("whereas our people are extremely badly off, with no houses to live in, a new palace should not be constructed"), Emperor Showa and Empress Kojun resided in Gobunko for many years, even after the war. It was in 1962, 16 years after the end of the war, that the "Fukiage Palace" was constructed next to Gobunko as a new residence. A further new Palace was constructed in the remains of the Meiji Palace 7 years later. After the demise of Emperor Showa, the Fukiage Palace was renamed "Fukiage Omiya Palace," for the purpose of using it as the residence of Empress Kojun. The present Emperor, Akihito (to avoid any confusion, hereafter referred to as Emperor Akihito), had resided in the Akasaka Palace (the current Crown Prince's Palace), moving back and forth between that palace and the Imperial Palace for a while after his enthronement. After then, he moved to a new residence in the Fukiage Garden constructed in 1993.

The old garden, which had previously been situated in the current Fukiage Garden and became dilapidated during the Meiji Restoration (1867–1877), was reconstructed into a more magnificent one when Emperor Meiji took up residence in the Meiji Palace. A half golf course (9 holes) was also constructed, where Emperor Showa and Empress Kojun enjoyed playing golf together before the war, as mentioned earlier.

Emperor Showa and Empress Kojun stopped using the golf course in 1937, when the Sino-Japanese War broke out. Afterwards, the plants in the garden were allowed to grow naturally, as maintenance had been discontinued at the Emperor's direction. Since 1948, following the end of the war, the plants growing naturally in Musashino were actively transplanted into the garden, with the aim of reproducing the natural environment of Musashino. These plants subsequently disappeared from Musashino one after another, due to overdevelopment during the high economic post-war growth period.

According to the five-year investigation conducted by the National Museum of Nature and Science starting in 1996, following the suggestion of Emperor Akihito, two kinds of woods were formed: one was made up of broad-leaved evergreen trees, such as *Ilex integra* and *Castanopsis sieboldii*, which were mainly planted on the periphery of the Fukiage Garden, and the other was made up of broad-leaved deciduous trees, such as sawtoothed oaks (*Quercus acutissima*), planted in the central area of the garden where there had been an open space in the past. The former woods were composed of plant species similar to those of an unspoiled grove, and in the latter, trees were planted for landscape gardening. An old pine tree, called "Dokan matsu," had previously stood in the garden. This name is associated with the story that it was planted by Dokan Ota (a samurai general) in the 15th century, but, in fact, an investigation of its annual growth ring conducted when it fell down revealed that its age was about 270 years, meaning it was younger than the long-accepted assumption.

Examples of the old trees that have survived from the Edo Period until recent years include a huge Japanese *Zelkova serrata* planted in an era (from the end of the 17th century to the beginning of the 18th century) called the Genroku Period of the Edo Period or later, and several pine trees collected from all over the country by the 11th Shogun, Ienari Tokugawa (1773–1841). Thus, the existing virgin forest-like woods lying on the grounds of the Imperial Palace have been formed out of a mixed combination of old trees, plants planted in recent years, plants naturally grown from seeds, and others. The Fukiage Garden, in which a fruit farm, fields, a rose garden and a mulberry field for silkworm cultivation together create an atmosphere of the village forest "Satoyama," sustains its diversity because of the many kinds of animals and plants that grow there, despite being just a narrow space in the center of Tokyo. In fact, raccoon dogs, masked palm-civets, American bullfrogs, snakes and goshawks live there, and about 100 species of myxomycetes, in which, as mentioned, Emperor Showa was especially interested, have been found (quoted from books, including "Organisms Living in the Imperial Palace and Fukiage Garden," [57] edited by the Research Group of Organisms in the Imperial Palace, the National Museum of Nature and Science).

The grounds of the Akasaka Detached Palace, where the Kamiyashiki (the feudal lord's residence during his term of Sankinkotai (alternate attendance in Edo)) of the Kishu Tokugawa family existed in the past, are adjacent, in the northeast, to the State Guest House (the old Akasaka Detached Palace) called the Geihinkan, constructed as the Crown Prince's Palace in the Meiji Period. The Geihinkan now serves as a guest house for VIPs from abroad. The residence of Emperor Meiji was temporarily situated on the grounds of the Akasaka Detached Palace after the Nishi-no-maru Palace in the Edo Castle was destroyed by a fire. Emperor Showa and Empress Kojun also briefly resided in the Akasaka Detached Palace for a period of time after getting married. At present, the Crown Prince's Palace (the old Akasaka Imperial House, where the Emperor Akihito resided), where the Crown Prince and Princess reside now, the residence of Prince Akishino (the old residence of Prince Chichibu), the residence of Prince Mikasa, the residence of Prince Takamatsu, and the eastern residence of Prince Mikasa (the old residence of Prince Mikasa Tomohito) stand on the grounds of the Akasaka Detached Palace amidst its vast green-rich environment. In the space occupying the central area of the grounds of the Akasaka Detached Palace, a pond surrounded by a circuit-style garden, called the Akasaka Imperial Gardens, has been formed. An Imperial Garden Party has been held in the garden twice every year, in spring and autumn, under the auspice of the Emperor since 1963. Many have likely seen scenes of the party on TV. In the past, capybaras, a sort of large-sized rodent native to South America, have often been kept as pets and, at present, several free-ranging Patagonian maras, closely related to but a little smaller than capybaras, are being kept in the garden by Prince Akishino's family. Prince Akishino and Princess Kiko have planted various kinds of plant in the garden. It is said that Princess Aiko, the first daughter of the Crown Prince, and Prince Hisahito, the first son of Prince Akishino, who are their next-door neighbors, enjoy frog- and insect-catching together.

Meanwhile, the term "Goyotei" means the villas of the Emperors and Imperial Families and, at present, three Imperial Villas exist: Hayama Imperial Villa, Nasu Imperial Villa and Suzaki Imperial Villa. In the past, Imperial Villas also existed in Atami, Hakone Miyanoshita, Nikko Tamozawa and Shiobara, including the Numazu Imperial Villa, which is close to the shore where Emperor Showa used to gather shells in his boyhood days. The Imperial Family members sometimes go to these villas and fully enjoy the nature that is specific to the local areas as a way of curing the fatigue associated with their myriad official duties. It is said that one of these villas, the Hayama Imperial Villa situated at Hayama-machi, Kanagawa, was constructed as a health resort for Emperor Meiji on the suggestion of Erwin von Bälz (1489–1913), a German court physician. Hayama was originally nothing but a rural fishing village, but, with the construction of the Imperial Villa in 1894 as a starting point, the statesmen who contributed to the Meiji Period, called the "Genkun of Meiji," built their villas around the Imperial Villa. In the Imperial Villa facing Isshiki Beach, the Imperial Family members have experienced joyful rest at the seashore for generations. They say that Emperor Taisho was pleased with the villa and often went there to rest. The Hayama Imperial Villa was Emperor Showa's favorite, where he was not only able to become absorbed in swimming and gathering animals and plants on the beach together with Empress Kojun and his children, but he also used it as a base for collecting marine organisms from Sagami Bay. This villa might also have provided Emperor Akihito with motivation toward research on fish belonging to the suborder Gobioidei, and he sometimes enjoys boating with his grandchildren on the sea nearby using the Japanese-style boat, as described earlier. The remains of the attached residence of the Hayama Imperial Villa, where Emperor Taisho passed away, is now open to the public as "Shiosai (sound of the sea) Park."

The Nasu Imperial Villa was constructed in 1926 in the highland area at an altitude of about 700 meters in Nasuno-ga-hara, which abounds with a wonderful natural environment, including a beech forest. The area of the existing site is 6,620,000 m^2, and the area of the old site, including the land the operation of which was transferred to the Ministry of the Environment for nature observation as "Nasu Heisei no Mori," was 12,220,000 m^2. This transfer of land operation was decided at Emperor Akihito's discretion. The Nasu Imperial Villa is the first villa to have been constructed as a resort house for the then-Prince Regent Hirohito, just married to then-Princess Nagako, and immediately after its construction, the Prince and the Princess stayed there for one month. I would guess that Emperor Showa had been pleased with the natural environment of Nasu when he visited there at an earlier time. Afterwards, as described earlier, the Emperor and the Empress visited the villa every year until the year before he passed away, with the exception of temporary discontinuation during the war. It is said that a golf course lay on the grounds of the villa in the early days of the Showa Period. At the Nasu Imperial Villa, Emperor Showa conducted nature observation and plant collection with Empress Kojun or his chamberlains and the members of the Biological Institute of the Imperial Household, sometimes throughout the day. Emperor Akihito's family and the Imperial family members often visited the Nasu Imperial Villa to rest and enjoy the nature-rich environment at Nasu. Crown Prince Naruhito and Crown Princess Masako, who like mountain-climbing, have often visited Nasu and have climbed Nasu Range. In recent years, Princess Aiko and Prince Hisahito have enjoyed insect-catching together (Fig. 2.2).

In 1933, Takanaga Mitsui, the head of Mitsui Zaibatsu (Combine), constructed the "Mitsui Institute of Marine Biology" on his vast villa site at Suzaki, Shimoda, Izu Peninsula, at his own expense. At the institute, young, prominent contemporary researchers, including Juro Ishida, an expert on biochemical embryology, and

Fig. 2.2 A bird's-eye view of the Nasu Imperial Villa (Courtesy of the Asahi Shimbun Company)

Hitoshi Enami, a researcher on the hormones of invertebrates, had met and actively developed their researches together. Unfortunately, the institute was used less after the war, and was closed when Takanobu, the first son of Takanaga, passed away in 1933, and the vast site of the institute was handed over to the Imperial Household. Since the Numazu Imperial Villa had gone out of use in 1969 and the main residence of the Hayama Imperial Villa was destroyed by arson in 1972, the Suzaki Imperial Villa was constructed in the autumn of that year. The villa has a private beach. As described earlier, Emperor Showa conducted research on the flora in the area around the beach and collected marine organisms offshore and from around its neighborhood. It is said that Emperor Akihito enjoys diving there.

As can be easily ascertained from the descriptions above, the Fukiage Garden, which is the residence of the Emperor and the Empress, the grounds of the Akasaka Detached Palace where the Imperial Family members reside and the individual villas are all situated in a nature-rich environment; therefore, Emperor Showa and Empress Kojun, as well as other members of the Imperial Family, had a great many opportunities to observe animals and plants since their childhood. It is unlikely that the children of the Imperial Families, who have grown up in a home that has produced biologists for three generations, would not be influenced by such a background. It is clear that their living environment is far richer in nature than those of the average ordinary folks who live in cities, in housing complexes, and who have no choice but to play in amusement parks, artificially constructed parks or school playgrounds.

The Imperial Family and Horses

Now, I would like to change to the topic of horses, in which the Imperial Family members have taken a great interest, and the Imperial Stock Farms. Horse bones and harnesses have been discovered during excavation of ancient burial mounds constructed in the 4th to 5th centuries in various places in Japan. "Gishi-wajin-den," a Chinese history book written at the end of the 3rd century, indicates that no horses lived in Wakoku (the name by which China (Sei-shin referred to Japan at that time), suggesting that it was directly before or after the 3rd century that horses were introduced to Japan. It is said that a legend has been passed down through the ages that says that the 1st Emperor Jinmu, while enjoying fishing in the vicinity of Nichinan-shi, Miyazaki, was given a horse named "Tatsuishi" by an old man, and the Emperor subsequently took care of it.

Stories indicating that horses were used in the eastern expedition of Prince Takeru Yamato and that a cavalry took part in the Jinshin War in 672 have also been passed down. These stories suggest that those at the highest levels of leadership, including the Emperor, sat astride their horses as a symbol of their power.

In the Nara (710–794) and Heian (794–1185) Periods, the games "Kurabe-uma" (traditional horse racing) and "Uma-yumi" (horseback archery) had come into fashion at the Imperial Court and other places. The former is similar to contemporary horse racing. It is said that the latter first appears in the description in "Nihon-shoki" (the oldest chronicles of Japan) of an instance in which the 21st Emperor Yuryaku invited a prince, who was a powerful candidate to become the next Emperor, to go hunting, and then shot him dead before his enthronement. The vehicle for nobles in the Heian Period was an ox carriage, while "The Tale of Genji" describes scenes in which young Hikaru Genji and Kaoru, who are characters in the book, mount their horses. Unfavorable stories have been passed down that the 57th Emperor Yozei (period of reign from 878 to 884), who often made the chamberlains extremely worried in regard to his immorality, and the 65th Emperor Kazan (period of reign from 984 to 986), who also behaved in an eccentric fashion, both rode their horses quite roughly around the Imperial Court. Both Emperors were forced to abdicate the throne in favor of the next Emperor under the Fujiwara clan's power (quoted from the book written by Hidehiko Kasahara [34]). All of these stories reflect the fact that the Emperors actually rode horses. Presents to the Imperial Court have included the horses raised at the Imperial Stock Farms situated in the Shinano (the current Nagano), Kanto and Tohoku regions. After administrative power had shifted from the noble class in the Heian Period to the warrior (samurai) class, such as the Minamoto and Taira clans, horses became indispensable for wars in the Kamakura (1185–1333), Muromachi (1336–1573) and Sengoku Periods (1467–1590).

Looking outside of the country, in the National Archeological Museum, Naples, Italy, where I studied half a century ago, one can find a mosaic of Alexander the Great fighting bravely on an extraordinary, unmanageable horse "Bucephalus." There are famous stories about Genghis Khan and his clan skillfully controlling

Mongolian horses during their successful conquest of the Eurasian Continent, and the portrait of Napoleon on a white horse titled "Napoleon Crossing the Alps," by Jacques-Louis David is also well known. For some reason, we have always enjoyed images of supreme rulers on their horses.

The royal families and nobles in the European countries routinely put horses to a variety of practical uses, for example, for sports and holding events such as riding, pulling carriages, racing, polo, hunting, etc. Similarly, the British Royal Family members have always been closely associated with horses, and Queen Elizabeth II herself also enjoys riding. Princess Anne, the first daughter of Elizabeth II, took part in the equestrian event at the 1976 Montreal Olympic Games, and Zara Philips, the grandchild of Elizabeth II, took part in the London Olympic Games, as one of the members of a team event and won the silver medal. Moreover, Mark Philips, Zara's father and Anne's ex-husband, has won both silver and gold medals in the Olympic Games. The successive British Royal Families have all been known for their patronage of horse racing, and Queen Elizabeth II is, of course, an enthusiastic patron, her horses having won victories in multiple stakes and she having frequently been chosen for the annual Leading Owner Award. The Ascot Racecourse, relatively close to Windsor Castle, is owned by the British Royal Family, and the family holds the Royal Ascot Race Meeting there every June. "Queen Elizabeth II" horse races have been held all over the world, including the Japanese "Queen Elizabeth Cup."

Perhaps it was knowledge of the European and US situations described above that caused Tomomi Iwakura and Sanetomi Sanjyo (leading court nobles in the Meiji Restoration) to suggest to the young Emporor Meiji that he should take horseback riding lessons after the Meiji Restoration. The Emperor studied and mastered Japanese-style horseback riding under the coaching of Tadatsuna Toda, and eventually created his own way of riding a horse. The painting "Oshu Junko Bahitsu Goran" ("The Emperor Meiji Observing Horses While Passing through the Tohoku District") is hung on the wall as one of a series of wall paintings displayed in the Meiji Memorial Picture Gallery, situated in the outer gardens of the Meiji Jingu Shrine. "Oshu Junko Bahitsu Goran" depicts the scene when the Emperor saw hundreds of horses at the precincts of Morioka Hachiman Shrine, Iwate, where he stopped while riding through Tohoku in 1876. He took particular note of a chestnut horse named "Kinkazan." "Kinkazan" was a fine horse, who ended up serving with the Emperor in many of his official duties, in roughly 130 different instances until the horse's death at the age of 26 in 1895. Kinkazan was then stuffed at the Emperor's instruction, and also now rests in the Meiji Memorial Picture Gallery.

It is said that Emperor Taisho, Emperor Meiji's successor to the Imperial throne, despite often appearing to be poor in health, especially as he grew older, continued to excel at horseback riding. This was likely because he had learned it with such enthusiasm in his student days in Gakushuin, and thus, in his later days, in addition to tasting wines and smoking, riding still served him as a way to relax.

Taka Adachi served Emperor Showa in his childhood as a nursing tutor, later marrying Kantaro Suzuki (an admiral), who served as the grand chamberlain and then as the prime minister at the end of the war. Taka recollected the many scenes in which Emperor Showa and Prince Chichibu rode the wooden horses that Emperor Meiji had presented to them, saying, "You will have many opportunities to ride horses in the future." Emperor Showa rode a native Korean horse that had been presented to him by Hirobumi Ito, Japan's first Prime Minister (the first governor-general of Korea) (quoted from the book written by Yukio Ito [29]). They say that the reason why he was so interested in living organisms in his childhood is that he was motivated considerably by Taka Adachi. As he enjoyed horseback riding so much, he began formal training in it around the fourth or fifth grade at Gakushuin Primary School, being coached by two experienced teachers, Toshu Nemura, the officer responsible for the nation's war horses, and Motoyoshi Mibu, a lieutenant colonel in the army cavalry, both of whom were general officials of the Imperial Household during almost all of the years that he studied at the Crown Prince's Imperial 'Study.' After the war, the Emperor was further trained under the tutelage of the former major general Kohei Yusa, acclaimed as a "Showa no Uma Syogun" ("expert equestrian of the Showa Period") or a "Showa no Heikuro Magaki" ("expert equestrian of the Edo Period") at the Palace Riding Club, situated in the East Gardens of the Imperial Palace. Kohei Yusa took part in the 9th Olympic Games in 1928 in Amsterdam as a rider and coach, and he subsequently took part in 6 more Olympic Games as a coach, up until the 1964 Tokyo Olympic Games, with the exception of the discontinuation during and after the war. He also contributed to the restoration of equestrian activities in Japan, where the Japanese military was demobilized and no military personnel were present. I still remember the brave figure of Kohei Yusa, with his Imperial moustache in his later years.

Most people usually remember the scene in which Emperor Showa mounts a white horse at a military review of the Imperial Japanese Army, but, as already described, before that, he also inspected the affected areas immediately after the Great Kanto Earthquake that happened in his Prince-Regent days while mounted on a horse. They say that the number of horses that he rode throughout his lifetime amounts to 50. After the war, he exhibited great puissance on the riding ground at the Imperial Palace.

However, he discontinued horseback riding in 1950. The white horses that he rode at military reviews included "Fubuki" ("Snowstorm") and "Shirayuki" ("White Snow"), Arabian horses native to Hungary (1925–1943), and "Hatsuyuki" ("First Snowfall"), an Anglo-Arab horse born in the Shimousa Imperial Stock Farm (1943–1945, the end of the war). The above-described Kohei Yusa was involved in the purchasing of the former two horses and the number of times Emperor Showa rode them are 254 and 344, respectively. After the end of the war, Emperor Akihito also rode "Hatsuyuki." Moreover, the Imperial Family members, including Emperors Showa and Akihito, and Prince Hitachi, rode Hatsuyuki's half-brothers, "Mineyuki" ("Snow on the Mountain Top") and "Hatsushimo" ("First Frost"). And, of course, the family also had several horses that were colors other than white.

Emperor Akihito also like horseback riding so much that he served as a captain of the equestrian club starting in his second grade year at Gakushuin High School and won the Kanto Equestrian High School Championship. Empress Michiko also enjoyed horseback riding. In addition, the Crown Prince and Crown Princess Masako have also enjoyed horseback riding, along with their other favorite hobby, mountain climbing. My father, a fellow student of Prince Chichibu's in his Gakushuin days, chose to become a military officer with his other fellow students when the Prince entered the Army Cadet School from the Gakushuin Junior High School, following the traditional practice of the Imperial Family at that time, and later became a cavalry officer. After the end of the war, he sometimes rode a horse in a Ouankai consisting of the ex-members of the Gakushuin Equestrian Club, and, at the age of about 80, served as a captain of the team of Emperor Akihito, who was the Crown Prince at the time, in a Dakyu (Japanese polo) game in the "New Year's First-Ride Athletic Meeting" hosted by the Ouankai. Prince Hiro Naruhito (the current Crown Prince) and Prince Aya Fumihito (the current Prince Akishino) both took part in this game, and, according to my memory, Prince Hiro's team won the match (Fig. 2.3). Emperor Akihito also liked show jumping, but, after his accession to the throne, he gave up it for fear that, if he were to be injured, his official duties might be disturbed. Tatsuhiko Kawashima, an economist and the father of Princess Kiko (Akishino), served as the professor in charge of the Gakushuin University Equestrian Club.

According to the materials of the Shumeryo (Bureau of Horses), the game that would come to be called Dakyu originated in one particular area of Central Asia and then propagated to Western countries, where it was renamed polo, while also

Fig. 2.3 A scene of a Dakyu game in the horseback riding field at Gakushuin. Emperor Akihito, the Crown Prince at that time, and the Imperial Prince Hiro Naruhito, the present Crown Prince, took part in the game (Supplied by the author)

propagating to China, where it was given its Asian name. It seems to have prop-agated to Japan via the Korean Peninsula in the 8th or 9th century. The Shumeryo houses the athletic equipment used in the game, of a style that was most prevalent during the Edo Period, although, since the Meiji Period, western style equestrianism has become prevalent, and the western style saddles used in polo are now also used in the Dakyu games hosted by Ouankai. Polo is a game in which two teams, each consisting of four members on horses, hit a ball into their opponent's goal using a stick. In the Dakyu game hosted by Ouankai, the red and white teams, each con-sisting of four or five members riding horses, scoop the ball up using a stick with a net attached to its top, called a Sade, similar to the stick used for lacrosse. Their objective is to get close to the opponent's goal while rotating the stick so that the centrifugal force prevents the ball from being dropped, and throw the ball into the goal. Empress Michiko, who learned the basics of the game from Emperor Akihito, also seemed to be good at Dakyu (quoted from the book written by Mototsugu Akashi, another one of the Emperor's fellow students [1]).

The Shumeryo, one bureau of the Imperial Household Agency, is responsible for the purchasing and training of horses, management of carriages and harnesses, operation of the Imperial Stock Farms, and other duties. Tracing back to old times, in the Nara Period, the Meryo was established out of the Hyobu-sho (Ministry of the Military) consisting of the Sameryo (Left Office) and Umeryo (Right Office), both of which were integrated together into one unit, the Shumeryo, in the Heian Period. This bureau was abolished after World War II and has now been reorga-nized into the Shume Section, Vehicles and Horses Division, Maintenance and Works Department. More than 30 horses for riding and pulling carriages are kept in the Imperial Palace alone. The farms, called the Imperial Stock Farms, are operated under the jurisdiction of the Imperial Household Agency. The Imperial Stock Farms were originated from Sakura-no-maki, situated in Shimousa (the current Chiba), in the Edo Period. On the suggestion of Toshimichi Okubo (a leader of the Meiji Restoration) to the Throne, a sheep farm and a livestock breeding station were constructed in Shimuosa in 1875 and, shortly after, reorganized into one unit, the Shimousa Livestock Breeding Station. After then, it was renamed the Shimousa (Sanrizuka) Imperial Stock Farms under the jurisdiction of the Imperial Household Agency. In 1969, as New Tokyo (Narita) International Airport was being con-structed, the Farms were transferred to Takanezawa-machi, Tochigi.

The Imperial Stock Farms produce riding and draft horses for the Imperial Family, as well as food, including beef, mutton or lamb, pork and chicken meat, and organic vegetables, which are used as ingredients for meals for the Imperial Family, and also for events such as banquets and garden parties. When the 2011 Great East Japan Earthquake occurred, the Imperial Stock Farms were damaged, just like other private houses and public facilities, but, in accordance with the intention of Emperor Akihito and Empress Michiko, the food from there was provided to the evacuation places as relief supplies after having undergone radioactivity inspection. Before the war, two thoroughbred stallions, Tournesol and Diolite, were imported to Sanrizuka from the UK, and they subsequently produced Wakataka, the winning horse at the first Tokyo Yushun Great Race, and St. Lite, who won the Triple Crown. I myself visited Sanritzuka, at which my paternal uncle worked, in my childhood and actually saw both of these horses. While not having as steadfast a relationship with horseracing as the British Royal Family, the Japanese Imperial Family have also been involved in racing; for example, the Tenno Sho (Emperor's Cup) Horse Race has been held in spring and autumn and Emperor Akihito and Empress Michiko have often been in attendance to watch the races.

Besides the horse races, the Imperial Household hosts a traditional event, duck hunting, on the hunting grounds at Ichikawa, Chiba and Koshigaya, Saitama, as recreation for the state guests who are visiting Japan. In the past, the ducks caught were immediately roasted on the hunting ground to eat, but, after Buichi Oishi, the Director General of the Environment Agency at that time, declined the invitation to the duck hunting event in the name of wildlife protection, a new tradition sprang up since 1973 in which the ducks caught are released with individually attached labels. At present, farm-raised Aigamos (crossbreeds between mallards and domestic ducks) are supplied for food instead of ducks, and ducks are protected as subjects for bird-watching. There is no doubt that the above-mentioned deep relationship with farm animals, including horses and wild birds, with which they have been familiar from their childhood, as well as the natural environments of the Imperial Palace and Villas, is partially responsible for the Imperial Family members being interested in animals and plants.

The Departments of Zoology and Botany, Faculty of Science, the University of Tokyo

Emperor Showa, as a specialist in the taxonomy of hydrozoans, collected a wide variety of marine organisms, Emperor Akihito is also an expert on taxonomy, in his case, of fish, and Prince Akishino has studied zoology at the University of Oxford and has conducted research on catfish and fowl. Accordingly, many zoologists have assisted the Imperial Family's researches. Most of these zoologists graduated from the Department of Zoology or the Department of Botany, Faculty of Science, University of Tokyo, because the Departments of Zoology and Botany did not exist at other universities, including Kyoto University, until 1919. As described earlier, Hirotaro Hattori, who had encouraged Emperor Showa in his biological research since his days at the Crown Prince's Imperial 'Study,' also graduated from the Department of Botany, University of Tokyo. Tomitaro Makino, Masaji Honda and Hiroshi Hara working at the Department of Botany, University of Tokyo, also supported the Emperor Showa's botanical research, in addition to Hattori. And many biologists contributed to the cultivation of plants in the wild plant garden of the Imperial Palace and the plant collection at Nasu and Suzaki.

Now, I would like to discuss the Departments of both Zoology and Botany as part of the organization of the University of Tokyo in those days. Incidentally, at present, these Departments have been integrated into the Department of Biological Science, Faculty of Science, University of Tokyo, together with the former Department of Anthropology. With respect to the graduate school, both the Department of Biophysics and Biochemistry and the Department of Bioinformatics and Systems Biology are now parts of the Department of Biological Science, Graduate School of Science, University of Tokyo. Accordingly, the former Departments are referred to in this book as the Departments of Zoology and Botany, respectively, so as to be easy to understand for the readers.

The Department of Biology was created in the Faculty of Science in 1899, when the University of Tokyo was founded. There were originally courses in both zoology and botany, but the students were only allowed to selectively major in either of them in the last year of the three-year period of attendance at the university. As described earlier, at that time, foreign teachers (called hired foreign teachers) were invited to Japan to deliver their lectures, including Edward Sylvester Morse, an American zoologist who was the first professor of the Department of Zoology. The second professor was Charles Otis Whitman, also an American zoologist. The third professor was Kakichi Mitsukuri, the first Japanese professor, who had studied at Yale University and Johns Hopkins University in the US. Meanwhile, as the first professor of the Department of Botany, Ryokichi Yatabe was appointed, in light of the nation's long-standing tradition of herbalism. He had studied at Cornell University in the US and then worked at the Tokyo Kaisei Gakko (Kaisei School), the predecessor of the University of Tokyo.

Morse, one of the disciples of Jean Louis Rodolphe Agassiz, advocated the need for sample collection to his students and encouraged them to make field observations, following the lesson taught by Agassiz, who put forth the famous words, "Study nature, not books." Morse actually took his students on a tour throughout Japan, from Hokkaido to Kyushu, for sample collection. Morse also made a huge contribution to the construction of a museum, launched the Japanese Society of Biology and strived to popularize the theory of evolution. Whitman, one of the disciples of Rudolf Leuckart, who was a prominent German zoologist, introduced the latest European and American technologies, such as Zeiss microscopes and microtomes, into Japan, and assigned different species of animals to his individual students as their research subjects. Unfortunately, the museum was abolished soon afterwards, forcing zoology in Japan to follow an ill-grown developmental path, as if it was a floating weed, without the sort of well-organized sample collections seen in European countries and the US.

Kakichi Mitsukuri, born into the family of a well-known Rangakusya (Dutch scholar), Shuhei Mitsukuri, had an elder brother, Dairoku Kikuchi, a mathematician who was the president of the University of Tokyo and then served as the Minister of Education, and a younger brother, Genpachi Mitsukuri, a historian, who was the author of "The History of the Great French Revolution" and "A Lecture on Western World History." As described earlier, Genpachi originally studied at the Department of Zoology. Kakichi Mitsukuri had an interest in the development of animals and made his students systematically investigate the development of the different animals in which each of them was interested. He also conducted researches into seeds, leading to the newly developed fields of genetics, physiology and ecology. Amongst these superior achievements, his greatest contribution is that he established Japan's first marine biological station at Misaki, Kanagawa, facing onto Sagami Bay, a treasury of marine organisms, in 1889. The date of its establishment was a little later than those of the well-known Stazione Zoologica di Napoli in Italy and the Roscoff Marine Biological Station in France, but earlier than those of the Marine Biological Laboratory, Woods Hole, in the US, and the Plymouth Marine Biological Station in the UK. This promoted the rapid progress of zoology in Japan. As already described, Emperor Showa visited this station as needed. In 1889, the Department of Biology was separated into the Departments of Zoology and Botany.

As a successor of Mitsukuri, Isao Iijima, who studied under Morse and Whitman and was a member of the 1st graduating class in 1881, was appointed as the second Japanese professor. He studied various kinds of animal, especially parasites and birds, under Rudolf Leuckart. Both Mitsukuri and Iijima, who were eminent, highly qualified professors, fostered many young biologists. One of these, Shozaburo Watase, a member of the 4th graduating class of the special course, stayed in the US for a long time and became a world authority on cell biology, ranked with Wilson. After his return to Japan, he served as the third professor at the Department of Zoology. He introduced new academic fields, such as applied zoology and biogeography, into Japan, but, after his death, these fields disappeared from the Department. Asajiro Oka, a member of the 8th graduating class of the special course (1887) and the author of "A Lecture on the Theory of Evolution," contributed to the dissemination of this theory and served as a founder of a human network stretching from Tokyo Higher Normal School to the Tokyo University of Arts and Sciences (later the Tokyo University of Education, and then Tsukuba University). Chujiro Sasaki, a member of the 1st graduating class, like Iijima, and Chiyomatsu Ishikawa, a member of the 2nd graduating class, contributed to the Departments of Entomology and Fisheries, respectively, at the Faculty of Agriculture, University of Tokyo. Chujiro Sasaki is the eponym of *Sasakia*, which is the generic name of a giant purple butterfly, the national butterfly of Japan. Ishikawa, who contributed to the dissemination of the theory of evolution, was also appointed as the Director of the Ueno Zoo, but just one year after his assumption of office, he tendered his resignation, because he had purchased a giraffe without permission of the Imperial Household Department, the competent authority at that time.

Once the Department of Zoology had been created at the Faculty of Sciences, Kyoto University, Tamiji Kawamura and Taku Komai, both of whom were transferred from the University of Tokyo, established foundations of animal ecology and ethology and zoological genetics, respectively. Yo Okada, who later returned to the University of Tokyo, created a pathway for experimental embryology. Against the background of the policy of increasing national prosperity and military power, biology was considered to be non-essential and non-urgent, and only three laboratories were created at the Department of Zoology, University of Tokyo, a situation that continued for a long time, until the post-war era. The researchers sought places other than the University of Tokyo for further development of their own researches. Specifically, zoological researches were conducted separately at the Departments of Zoology, University of Tokyo and Kyoto University, and the Departments of Agriculture, Veterinary and Animal Husbandry and Fisheries, Faculty of Agriculture, University of Tokyo. For this reason, it could probably be said that these separate researches formed the combined total zoological field in Japan at that time. Later, other former imperial universities and the Tokyo University of Arts and Sciences added to this group. At Hokkaido University, entomology had traditionally been developed under the guidance of Shonen Matsumura, called "the father of entomology" in Japan, starting at the Sapporo Agricultural School. Shonen Matsumura was the author of "A Thousand Insects of Japan," which Emperor Showa was fond of reading when he was a child.

Unfortunately, Mitsukuri died at the age of 51. After his death, Seitaro Goto followed in his footsteps and Naohide Yatsu succeeded Iijima in his post. While visiting foreign institutes in Europe and the US to learn closely how researches were conducted there, Yatsu saw with his own eyes that the researchers there not only observed and recorded animals, but also carried out experiments on them. Moreover, he himself did research on experimental embryology in the US and Naples. He contributed to the dissemination of laboratory animal science in Japan, as is clear from the episode, from around 1930, in which he stated firmly in front of the students at the Department of Zoology, "Forget everything about taxonomy that you have learned before this class." Both Professors Iijima and Goto, who were Yatsu's predecessors, had opposed any experiments in the course of which natural objects were artificially processed. Accordingly, from what I have heard, Yatsu, who was appointed as an assistant professor at the end of the Meiji period, experienced many difficulties before he attained the position of professor. Nevertheless, he had a deep knowledge of animal taxonomy, and wrote "Dobutsu Bunrui Hyo" ("The Animal Taxonomy List"), which was a handbook for students learning zoology both during and after the war. I and my colleagues also used this handbook to great effect in our student days. The handbook was very unique, in that it explained the origins of the names of individual animals, along with other useful information and anecdotes about them, entries that were pleasing simply to look at, and because one of its double pages was left blank for the purpose of taking original notes. Like Yatsu, Kiyoshi Takewaki, who was a researcher on endocrinology and attained the position of a professor after the war, had a deep knowledge of taxonomy.

Since then, the elements of natural history have been excluded, except for fish taxonomy, described later, and thus physiology, in a broad sense, has dominated the Department of Zoology, University of Tokyo, up to the present time. With a limited number of laboratories, this change may have been necessary in order to keep up with the global scientific trends. However, both Yatsu's merits and his faults, which had an influence on zoology in Japan, are debatable today. Toru Uchida and Teizo Esaki took over the tradition of taxonomy at Hokkaido University and Kyushu University, respectively, leading the Japanese world of natural history in the good old days. Unfortunately, a declination of the study of natural history at its original home, the University of Tokyo, inevitably exerted a substantial influence on the Japanese field of zoology (Cf. Mohri and Yasugi [45]). Emperor Showa stated his view on the situation as such:

> Taxonomy was abandoned, even at the University of Tokyo, the first-ranked university in Japan, while somehow being continued at Kyoto University. It is a real disappointment (quoted from the book by Munetetsu Tei [61]).

Meanwhile, perhaps relevant to the selection of zoology by Emperor Showa, many of the members of the Imperial family and noble families were enrolled in the Department of Zoology and made researches related mainly to natural history as so-called "noble biologists" [30]. One of these biologists was Yoshimaro Yamashina, who was an elder cousin by one year of Emperor Showa and was born as the second son to Prince Yamashina's family (Fig. 2.4). At that time, there were too many houses of Imperial princes, and he was demoted from the Imperial Family to the status of a subject, subsequently establishing the marquis Yamashina family. He worked in the army for five years, and then entered the University of Tokyo to study in a special course for two years. He founded the Yamashina Institute for Ornithology, which will be described in detail later in association with Sayako Kuroda (the former Princess). Along with Yamashina, the researchers on birds include Duke Nobusuke Takatsukasa, a member of one of the Gosekke (the five regent houses), Marquis Nagamichi Kuroda and his son, Nagahisa, in the family line of the former lord of the Fukuoka Domain, and Count Yukiyasu Kiyosu, a member of the family branch of Prince Fushimi, while Viscount Masanari Takagi, who was the father of Princess Yuriko, the wife of Prince Mikasa, was a researcher on insects. He was a brother-in-law of grand chamberlain Sukemasa Irie. Katsura Oyama, a compiler of "The Sea Shells of Sagami Bay," was a grandchild of Duke Iwao Oyama, a marshal of the Imperial Japanese Army. Many nobles, whose major fields were not always natural history, studied at the Department of Zoology, including Tadao Sato (his original family name was Kuroda) and Katsuma Dan. In contrast, a smaller number of nobles learned at the Department of Botany, including Yoshichika Tokugawa, the lord of Owari Domain, who established Japan's first privately-owned "Tokugawa Biological Institute," making a great contribution to the promotion of biology, and Baron Hiroshi Hara, a botanical taxonomist, whose father served as the president of the Privy Council (A Family of Noble Biologists [30]). Unique among these ex-students, Senji Yamamoto, a member of the Diet (Japan Labour-Farmer Party), also graduated from the Department of Zoology in 1920; he was eventually assassinated on the grounds that he had opposed the revision of the Maintenance of Public Order Act.

These noble biologists had made efforts to support natural history and scientific museums, which, unfortunately, had not taken root in Japan, by means of their plentiful funds, until the peerage system was abandoned, and, afterward, through their passion and sense of responsibility towards academics. They had put their hearts into these activities since the courses of taxonomy and ecology had been abolished, mainly because Yatsu had introduced experimental zoology, and the Japanese Government, which was completely devoted to the wealth and military

Fig. 2.4 Yoshimaro
Yamashina, the founder of the
Yamashina Institute for
Ornithology (Courtesy of the
Yamashina Institute for
Ornithology)

strength of Japan in those days, was unconcerned about the situation in the academic fields.

Many students from high schools under the old systems of education were disappointed to discover that these courses had been abolished from the Department of Zoology when they entered the University of Tokyo. Meanwhile, commoners tended to avoid selecting these academic fields, specializing instead in the frontier types of research that were coming to the forefront in those days under the assumption that only nobles and rich people would be willing to select these academic fields.

At last, after the war, two laboratories, one for Radiation Biology and one for Physiological Chemistry (Developmental Biochemistry), were added to the Department of Zoology, but the Laboratory of Genetics was limited only to the Department of Botany. Recently, the five laboratories were integrated into the Division of Zoological Science, with their research area essentially remaining the same as before. Against this background, it can be considered a step forward that the Division of Evolutionary Biology (a name no longer in use at the present time), including the staff of the National Museum of Nature and Science, was created in cooperation with the botanical researchers. The post-war professors of the Department of Zoology, University of Tokyo, including Yo Okada, gave lectures on the current state of zoology to Emperor Showa. In addition, lectures related to biology were also given to the Emperor by a wide range of biologists, from amateur researchers out of power, including Suguru Igarashi, a collector of butterflies who elucidated the life of the swallowtail, *Teinopalpus imperialis*, to the professors from the universities, including Taku Komai, a professor at Kyoto University, as described earlier.

On the other hand, the Department of Botany was formed by Japanese researchers alone, including Ryokichi Yatabe and Kensuke Ito, a botanist from the Owari Domain (the current Aichi), who studied under the direct guidance of von Siebold and who also contributed to the Department as an extraordinary professor. At the time when the Department of Botany was formed, the Koishikawa Medicinal Herb Garden operated by the Tokugawa shogunate was integrated into it as the Koishikawa Botanical Garden, and it has been open to the public since 1884. The Department of Zoology had been around on the Hongo campus for a long time, while the Department of Botany was situated in the Koishikawa Botanical Garden during the period from 1897 to 1934. It was one year before the transfer of the Department to the Botanical Garden that Goro Hirase found the spermatozoid of *Ginkgo biloba*, which is a well-known discovery to come out of his research. Thanks to the Department of Botany's historical genealogy of taxonomy and its convenient location, taxonomy has been traditionally taken over by the botanists there, including Tomitaro Makino, who studied under Yatabe; Takenoshin Nakai, who served as the director of the National Museum of Nature and Science; Hiroshi Hara, who was a disciple of Nakai; Masaji Honda; and Fumio Maekawa. The way for ecology was paved by Manabu Miyoshi (quoted from "The History of Botanical Researches in Japan—The 300-Year History of the Koishikawa Botanical Garden" [49]. Thus, a number of researchers who graduated from the Department of Botany were involved in the research conducted by Emperor Showa, as described earlier. Tetsuo Koyama, who explained the plants to the Emperor at the New York Botanical Garden, was one of these researchers. Yukio Yamada, a professor at Hokkaido University, who examined the seaweed collected by Emperor Showa and Empress Kojun from Sagami Bay, also graduated from the Department of Botany.

Generally, from the viewpoint of seeking samples and pursuing their studies further in various parts of the world, we feel that Kyoto University has dominated the zoological territory, while the University of Tokyo has been leading the botanical area, because of its long-standing laboratories dedicated to ecology and taxonomy. Needless to say, new academic areas, such as cytology and physiology, have been introduced into the Department of Botany, University of Tokyo. Fortunately, two laboratories, one a genetics laboratory and another a donation-funded laboratory, were added to the Department of Botany, which originally had three laboratories, as did the Department of Zoology, totaling up to five for the Botany Department in the Taisho Period. This sequence of events had a strong effect on the long-standing existence of the laboratories dedicated to taxonomy and ecology. After the war, the two laboratories were transferred to a newly-established Department of Biophysics and Biochemistry, and the Department of Botany was then re-organized into the Division of Plant Science, which was larger in scale. As described earlier, in 2014, the Department of Biophysics and Biochemistry and the Department of Bioinformatics and Systems were integrated into the Department of Biological Science, which was composed of zoology, botany and anthropology, forming the Department of Biological Science, Graduate School of Science.

The Personal History of Emperor Akihito

Now, we will describe how the present Emperor, Akihito (hereafter referred to as Emperor Akihito), became associated with biology. Those born in the same generation as myself perhaps recall a song that was popular at the beginning of the Showa Period (lyrics by Hakushu Kitahara and music by Shinpei Nakayama); "Natta natta po-o-po siren siren... kotaishi-sama oumare natta..." ("The siren sounds po-o-po, informing the nation that the Crown Prince has just been born..."). Crown Prince Tsugu Akihito, the present Emperor, was born as the first son of the Imperial Family, December 23rd, 1933. Emperor Showa and Empress Kojun had only had Princesses, four in total, Teru Shigeko (who later changed her family name to Higashikuni), Hisa Sachiko (who passed away 6 months after her birth), Taka Kazuko (who later changed her family name to Takatsukasa) and Yori Atsuko (who later changed her family name to Ikeda), until the Crown Prince was born. For this reason, the whole nation celebrated the long-awaited birth of the Crown Prince. I had long suffered under the misapprehension that the onomatopoeia "po-o-po" in the lyrics was inserted to keep the rhythm and ascribed it no meaning for a long time. In fact, it turns out that the sound represents a siren that informs us as to whether a baby that has been born is male or female; when the siren sounded once for one minute, the baby was female, and when it sounded twice with a ten-second rest between blasts, the baby was male. Two years after that, Prince Yoshi Masahito (the current Prince Hitachi) and, after a brief interval, Princess Suga Takako (who later changed her family name to Shimazu) were born.

In fact, as with Emperor Akihito and Empress Michiko, Emperor Showa and Empress Kojun initially intended to bring up their children themselves. It was natural that Emperor Showa desired to do so, because he spent his childhood with his parents, Emperor Taisho and Empress Teimei, in an outbuilding next to the Imperial House, almost like living in an ordinary home, except for the period of two or three years after his birth, during which he was nursed at the Kawamuras' house. Nevertheless, his desire was quashed at that time for the reason that the children could not be strictly brought up by their parents. Ultimately, the Princesses were moved to Kuretake-ryo, an outbuilding situated next to the Imperial House, one by one when they entered Gakushuin Girls' Primary School, as described earlier. The close advisors at the Imperial Court, including Nobuaki Makino, a Naidaijin (Minister of the Interior), and Kinmochi Saionji, a Genro, were similarly adamant in rejecting the wishes of the

Emperor and the Empress to bring up Crown Prince Akihito by themselves, at least until he entered Gakushuin Primary School. Accordingly, the Crown Prince was moved to the Temporary Crown Prince's Palace constructed in the Akasaka Detached Palace at the age of 3 years and 3 months, where he was raised by togu-fuikukans (tutors in charge of the Crown Prince) and court ladies in charge of nurturing him. It is said that only Kantaro Suzuki, the grand chamberlain at that time, disagreed with the opinion of the close advisors, and Saionji. Crown Prince Akihito, having been kept apart from his younger brother, Prince Hitachi, was raised in a stricter manner than Emperor Showa, who had lived with his younger brothers, Princes Chichibu and Takamatsu. The head tutor was Iwakichi Ishikawa, a professor at Kokugakuin University, who was later appointed as the chairman and then president of the university. Torahiko Nagazumi and Motofumi Higashisono, who served as the Shoten-cho (chief ritualist) were also tutors at one point. It has been said that Higashisono was adopted into the court-noble Higashisono Family from the Date Clan in Sendai; he also later served Prince Hitachi, and coached Crown Prince Akihito on how to ride a horse (Mototsugu Akashi [1]).

Crown Prince Akihito was reared during the turbulent times in which the Sino-Japanese War and the Pacific War occurred, and the post-war society was impoverished. He got kindergarten education with the children of the Gakushuin Girl's Primary School Affiliated Kindergarten, who later became his fellow students, at the former Biological Laboratory of the Imperial Household situated in Akasaka. The Crown Prince seemed to be very fond of dogs from childhood up. Taking the entrance of the Crown Prince into Gakushuin Primary School as an opportunity, Katsunoshin Yamanashi (Admiral of the Navy) was appointed as the President of Gakushuin. They say that Yamanashi, who made his best efforts towards disarmament of the navy during the period ranging from the Taisho Period to the Showa Period, had never used the familiar expression "Shinshu Fumetsu" ("Immortal Land of Gods"), which was often heard during the war. On the other hand, the Crown Prince needed to be educated so that he might develop the appropriate personality for an "Arahitogami" ("living god"). The fact that Moto Akiyama, a specialist in science and mathematics, took continuous charge of the "East class" to which the Crown Prince belonged might have influenced his future development into a biologist.

During the period toward the end of the war, the Crown Prince, as a pupil at Gakushuin Primary School, was evacuated, along with his fellow children, to the Numazu Imperial Villa, the Nikko Tamozawa Imperial Villa and Okunikko Yumoto to evade attacks by US Air Force bombers, as happened with all of the children living in urban areas. At the Tamozawa Imperial Villa, he got an education at the Nikko Botanical Garden, University of Tokyo, and apparently enjoyed fishing along the Daiya River. One of the Crown Prince's educators was Koichi Nomura, who had worked as a professor at and then retired from Gakushuin Boys' Senior High School. The successive Crown Princes had customarily been appointed as second lieutenants in the Army and Navy at the age of 10, but Crown Prince Akihito had no experience serving in the Army or Navy, because Emperor Showa refused to carry on such a tradition.

Meanwhile, Japan came under the occupation of the General Headquarters (GHQ) at the end of the war, and the trend of the times suddenly changed toward democracy. Honestly speaking, those who grew up as militaristic boys and girls during the war, including myself, were astonished at the drastic transformation in Japanese society, especially in the adults. However, there was nothing like finally being at peace. In the year following the end of the war, the Crown Prince entered junior high school. At that time, plans for the Crown Prince's Imperial 'Study', were scrapped, reflecting the GHQ's intentions. For this reason, the Crown Prince studied with many of his fellow students at Gakushuin Boys' Junior and Senior High Schools, unlike the Emperor Showa, who studied with a limited number of fellow students. It is a matter worthy of note that prominent scholars such as Tetsuji Morohashi, who was an editor of "Dai Kanwa Jiten" ("The Major Kanji Dictionary"), and Masao Kotani, a physicist, who were supposed to have served as lecturers at the 'Study,' delivered their lectures on kingcraft to the Crown Prince at Gakushuin.

Mrs. Elizabeth Janet Gray Vining was appointed as a private English tutor to the Crown Prince and Prince Hitachi by the GHQ upon the request of Emperor Showa, serving them over a period of a few years from the year following the end of the war (Fig. 2.5). Reginald Horace Blyth, who was a British teacher and taught English at both the former Forth High School and Gakushuin during the war, also delivered a lecture on "noblesse oblige" to the Crown Prince as his private tutor. Since 1949, Shinzo Koizumi, an economist who was the president of Keio University, had served the Crown Prince as an educator and delivered a lecture on

Fig. 2.5 Emperor Akihito reading an English book with Elizabeth Vining in 1948 (Courtesy of the Imperial Household Agency)

kingcraft that was appropriate for the Emperor as a symbol of the new era. It is guessed that the Crown Prince was profoundly affected by both Elizabeth Vining and Shinzo Koizumi in his boyhood days.

According to an article on the Crown Prince, then a second-year student at Gakushuin Boys' Junior High School, that appeared in an issue of Asahi Shimbun early in 1948:

> The Crown Prince is fond of history and biology, and is a voracious reader of such books as "The World History of Mankind" and magazines such as "Reader's Digest," as well as magazines dedicated to science. Among his hobbies, he excels at taking photographs and drawing pictures, and has just recently begun oil painting. He also likes animals, and takes personal care of cranes, rabbits, geese, dogs and Japanese medakas, while also conducting scientific research about them. For example, when a medaka had not laid any eggs, he conducted research on the reason why, and then made efforts to allow the eggs, once laid, to hatch.

The article goes on to mention that Emperor Akihito also became interested in history and biology in his childhood, like his father, Emperor Showa. He sometimes collected butterflies, together with Emperor Showa and Prince Hitachi, at Nasu. Shinya Inoué at the Marine Biological Laboratory, Woods Hole, stated, at the award ceremony of the International Prize for Biology (a prize that, as mentioned earlier, he had once won himself):

"It was shortly after the end of World War II that we welcomed to the Marine Biological Station in Misaki, Emperor Showa together with the Royal Family, including Your Majesty the Crown Prince, still in grade school, yet already fascinated by the behavior of electric eels" (quoted from Record of the 19th Award Ceremony of International Prize for Biology [22]). At that time, the Crown Prince was looking forward to going to the beach. Although he did not live with the Imperial Family, it is natural that the Crown Prince has become interested in biology, because he grew up in a nature-rich family environment, led by that eminent biologist Emperor Showa.

According to my own dim memory, it had been known to the public that Prince Hitachi, a younger brother of Emperor Akihito, was fond of living organisms. Whereas it was perhaps not until somewhat later that the Crown Prince's professional research on biology became known to the public, owing to the fact that his visits to foreign countries and his marriage to Michiko Shoda were so heavily covered. At the press conference on the occasion of his 70th birthday, Prince Hitachi stated:

> I went to Numazu after graduation from junior high school, and sometimes went to the sea, together with the Emperor, who was also fond of various kinds of fish, to enjoy catching stonefish and surf-fishing. At night, by flashlight, we went fishing for squid.

At that time, Emperor Akihito was maybe a senior in high school.

The Emperor engaged in many activities at the Gakushin Boy's Senior High School, including serving as captain of the equestrian club. Despite his interest in history and biology, he selected politics and economics as his majors when entering the university. One of the reasons for this was that there was no department or course at Gakushin University in which the students could study biology (at present, a Department of Life Science has been established), and another was perhaps the suggestion by Shinzo Koizumi that "Politics is essential to learn kingcraft. I recommend that you enjoy biology for pleasure." It seems that the Imperial Household Agency had the same opinion as Koizumi. Despite being in a different position from that of Emperor Showa, who gave up his desire to learn history, Emperor Akihito could also not simply select a course to follow at his own discretion. For this reason, Emperor Akihito seemed to be less interested in the lectures delivered at the Department of Political Science and Economics.

At the end of March 1953, the year after the Treaty of San Francisco came into force, which allowed Japan to free itself from Allied occupation, Crown Prince Akihito, a 19-year-old student at that time, went abroad for the first time to attend the coronation of Elizabeth II on behalf of Emperor Showa. It is said that Prince Chichibu, who had experience studying in the UK, strongly suggested that the Crown Prince attend the ceremony. The Crown Prince being about the same age as the Crown Prince Akihito in those days, the Emperor Showa made his visits to Europe by warship, while he had a one-month trip to London via the US and Canada by ship, airplane and railway.

Although, at that time, British citizens had a complicated sentiment toward Japan, due to ill feelings caused by World War II, his visit to London was thankfully devoid of troubles, owing to the fact that the war was over and thanks to sophisticated mediation by Prime Minister Winston Churchill. The Crown Prince stayed in the UK for 40 days and then visited eleven other countries, including France. In Europe, he met with King Baudouin of Belgium and Queen Margrethe II of Denmark. Ever since then, close relationships have been maintained between Japan and these countries. On his way back to Japan by airplane, he stopped in Washington DC, the US, to meet with President Eisenhower, successfully playing an important role in the interest of Imperial Diplomacy on behalf of the Emperor Showa, for whom traveling to foreign countries was difficult. He returned to Japan about 6 months after the start of his journey. This tour of foreign countries was extremely meaningful for both Japan itself and the Crown Prince Akihito. However, the extended foreign travel resulted in his having an insufficient number of credits, preventing him from moving up to the next year at Gakushuin University, which maintains a strict educational philosophy. Thus, he had no choice but to leave school and became a special student. He even quit the riding club (Motoaki Akashi [1]).

The Marriage of the Crown Prince and His Reign as Emperor (Heisei Period, 1989–2019)

In the spring of 1959, a gorgeous wedding ceremony and parade were held to celebrate the royal wedding of Crown Prince Akihito and Miss Michiko Shoda, a commoner, who had become acquainted at a tennis court situated in Karuizawa. At that time, great love for "Mitchi" (an informal name for Michiko) was sweeping Japan. In the following year, the first son, Prince Hiro Naruhito, was born. The second son, Prince Aya Fumihito, and Princess Nori Sayako were born in 1965 and 1969, respectively. Crown Prince Akihito, who had lived away from his parents and felt lonely when he was a child, and Princes Michiko took a stance of keeping their children at hand so as to bring them up themselves. For this reason, the traditional Togu-fuikukan (tutor-in-charge of the Crown Prince) system and the wet-nurse system were abolished. As the social situation had changed with time after the war, the Crown Prince Couple were able to enforce their will, unlike in the case of the Emperor Showa. The couple was committed to rearing their children fondly but strictly, while often making reciprocal visits to the foreign countries of state guests on behalf of the Emperor Showa. This type of official duty had gradually been increased. They visited 36 foreign countries in his Crown Prince days, for example, making two trips in the year when the Prince Naruhito was born, or visiting four countries in one trip, depending on the circumstances. They also visited all of the 47 prefectures in Japan.

After his marriage, Crown Prince Akihito began his taxonomical research on gobioid fishes (the suborder Gobioidei) in a room at the Crown Prince's Palace, as will be described later. When Emperor Akihito received the King Charles II Medal from the Royal Society, he stated in his acceptance speech:

In the 1960s, when I started my research on gobioid fishes, there were not yet many specialists in that field, and few up-to-date references.

According to the researchers working at the current Biological Laboratory, his first paper on gobies was published in 1963. His research required a vast amount of time, at least two to three years. It may be guessed that he had already started his research on gobies in 1960. Initially, he seemed to think that he might transfer the results obtained from his observations to another researcher for publication but he most likely warmed to his research over time, eventually merging the results into his own papers (Fig. 2.6).

In 1989, Emperor Showa passed away, and Crown Prince Akihito was enthroned at the age of 55, the name of the era thereby changing from Showa to Heisei. He acceded to the throne at the second oldest age among successive Emperors, next to the 49th Emperor Konin, who had assumed the throne after age 60, after Dokyo (figuratively referred to as Rasputin in Japan) had resigned from his position of central power and was sent to the Province of Shimotsuke (the current Tochigi) in 767. Following his important role in previous official duties, such as his foreign tour, his official presence at national events, including the National Tree-Planting Festival, had gradually been increased, in anticipation of the day when Emperor Showa would pass away. As described earlier, Emperor Akihito moved to the Imperial Palace from the Akasaka Detached Palace, which was used as the Crown Prince's Palace, for a while after his accession to the throne, and, in 1993, transferred to the current, at the time newly constructed Emperor's Palace. Emperor Akihito made great contributions to Imperial Diplomacy with Empress Michiko, by visiting China (1991), countries in North and South America, and European

Fig. 2.6 Emperor Akihito conducting his research in the Crown Prince's Palace when he was the Crown Prince (Courtesy of the Imperial Household Agency)

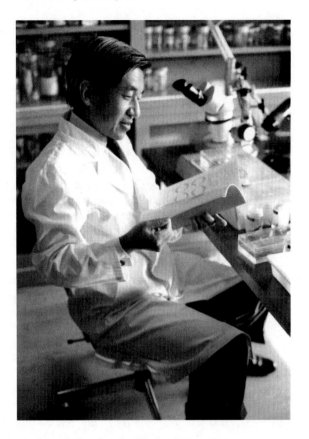

countries, and starting with three countries in Southeast Asia in 1990. They visited India in 2013 at the age of almost 80, and Belgium the following year.

During their revisit to the UK in 1998, the treatment of prisoners from the British Armed Forces during World War II by the Japanese Army (well known as a result of, for example, the scenes of the construction work on the Taimen Railway in the film "The Bridge on the River Kwai") was sharply criticized, though this matter had been superficially settled. In fact, a group of ex-prisoners protested their treatment by the former Japanese Army by turning their backs to the horse-drawn carriage carrying Emperor Akihito and Empress Michiko during the parade that took them to Buckingham Palace. This took place in the presence of Queen Elizabeth II and Prince Philip, Duke of Edinburgh, on the day following the Japanese Royal Couple's arrival in London on May 25th. Emperor Akihito stated, at a dinner party welcoming the Imperial Couple that was held at Buckingham Palace:

"I feel deep sorrow when I remember the great number of people who were deeply hurt by the war" (quoted from the book "Record of the Attendance of Crown Prince Akihito at the coronation of Elizabeth II" by Masaru Hatano [18]). As reports of the visit revealed that the Imperial Couple had displayed a sincere attitude, rather than simply giving superficial responses, and that there had been friendly interaction between the Princess Michiko and an English girl, the protest movement gradually abated, and the press altered their tone to reflect a more positive stance.

The same scenario was repeated when they visited the Netherlands two years later, in 2000. The Dutch people, as mentioned earlier, had a stronger sense of resentment against Japan because of the war. In fact, strong protest movements occurred when Emperor Showa and Empress Kojun visited the Netherlands. However, a long, deep silent prayer held by the Imperial Couple in front of a war memorial, combined with a sincere discussion with the representatives of the ex-prisoners, resulted in the former prime minister of the Netherlands sending a letter to the then-grand chamberlain Watanabe, which stated that the "Sincere personality of the Imperial Couple has changed the resentment and remoteness of Netherlanders in regard to Japan into great affection" (quoted from the book by Makoto Watanabe [68]).

This kind of conflict was no longer in evidence in 2012, when the Emperor Akihito attended Queen Elizabeth II's 69th coronation anniversary. As in the case of Emperor Showa, Emperor Akihito's dignity and sincere personality resulted in successful diplomacy, something that had eluded (and continues to elude) prime ministers and diplomats.

As previously mentioned, the Emperor Akihito, when he visited the UK on May 28th, 1998, was given the King Charles II Medal by the Royal Society. The Royal Society was formed by King Charles II more than 350 years ago. The medal named for him is meant to honor heads of states who have made great contributions to the promotion of science and technology. Emperor Akihito was the first recipient of this medal. It is not going too far to say that the medal was initially established to praise Emperor Akihito's highly-acclaimed research on gobies, conducted over many years. The only other winner of this medal, in 2007, was Abdul Kalam, the 11th President of the Republic of India and an aeronautical technologist.

In his acceptance speech, delivered at the award ceremony, Emperor Akihito modestly stated:

"When I was Crown Prince, I was elected a foreign member of the Linnaean Society of London. It was a great honour indeed to be admitted as one among the only 50 foreign members. While I felt that this was too great an honour for me, it stimulated me to continue my studies so as to maintain the high standards of research worthy of this honour, which remains one of my dearest memories.

Since my accession to the throne, my full daily schedule has so far prevented me from continuing my researches effectively. Indeed, the only paper I completed since enthronement was one which had been largely done some ten years before, when I was still Crown Prince. I am firmly resolved, however, not to let my light of learning be extinguished, I hope to go on pursuing my studies, mindful of the fact that I was today awarded the King Charles the Second Medal." (quoted from the book "The

Emperor Akihito Talks About Natural Science," edited by Asahi Shimbun [4]). As can be seen from the achievements of the Emperor Akihito that will be described later, he did indeed continue his research and published a number of papers.

The treatment of prisoners by the Japanese Army during the war was widely reported by the press, but the bestowal of this honorable award was given little coverage. Setsuro Ebashi, a winner of the International Prize for Biology and a former foreign member of the Royal Society, described introducing the King Charles II Medal in a magazine article:

"The press sensationalized the Emperor's visit to the UK this time, but only one magazine, *Bungeishunju* (September Special Issue), reported the significance of this honorable award" (quoted from the article by Setsuro Ebashi [11]).

Being the director general of the National Institute for Basic Biology, Okazaki, at that time, I really lamented such circumstances in Japan, sharing my disappointment with Kiyoshi Hama, an electron microscopist and the director general of the National Institute of Physiological Science.

It is unforgettable for us that not only had the Imperial Couple, who had wished to visit Saipan, made a tour to console the souls of war victims, but they were also filled with deep grief over the Battle of Okinawa, and showed heartfelt consideration for the victims of major disasters whenever they occurred. In 2007, the Imperial Couple visited Sweden for the first time in twenty-odd years, going to Uppsala with King Gustaf of Sweden and his Queen to attend the ceremony commemorating the 300th anniversary of Linné's birth. Emperor Akihito took his place among the honorary scholars of Uppsala University with the former Secretary-General of the United Nations, Kofi Atta Annan. Moreover, he delivered a keynote address in London upon the request of the Linnean Society. How proud the Japanese people are, to have Emperor Akihito of Japan, a country a great distance from Europe, invited by the society to deliver a keynote address (Fig. 2.7). Empress Michiko, for her part, wrote a waka poem at Uppsala:

The Emperor, himself a scientist, is staying at Uppsala to celebrate the 300th anniversary of Linné's birth.

From the aspect of biology, Emperor Akihito is not only interested in fish, but also has no less of a wide variety of knowledge of plants than Emperor Showa did. It is said that, when hiking together with the Empress Michiko and the little Prince Naruhito in those days, he always took pictorial guide books of flora with him to consult so as to confirm unknown species of plants, whenever they were found. A certain well-known scholar said that Emperor Akihito's knowledge of plants was equivalent to that of the professors at the University of Tokyo and Kyoto University.

Emperor Akihito wrote a waka poem at the 2001 Utakai Hajime (Imperial New Year's Poetry Reading) for which "Grass" was given as the motif:

Immersed in memories of the flowers that my father and mother cherished, I walk around the grassland at Nasu with the Empress.

Fig. 2.7 Emperor Akihito delivering his keynote speech at the event celebrating the 300th anniversary of Linnean birth (Courtesy of the Asahi Shimbun Company)

Thinking about the old days when she had taken lessons in plants from Emperor Showa, Empress Michiko wrote her own waka poem:

> I miss the old days, when, seeing flowers blooming in a marsh lying in the field at Nasu, Emperor Showa taught me that they were Hitsujigusa.

As you may know, Hitsujigusa (*Nymphaea tetragona*) is a kind of water lily that originated in Japan. The seal (equivalent to a family crest, imprinted on the belongings of individual Imperial Families) of Princess Sayako was a Hitsujigusa.

Empress Michiko has always understood and given deep consideration to Emperor Akihito's researches on gobies and other fish. She has not studied biology herself, but she does enjoy playing tennis, and even Dakyu, as well as being engaged in musical interests through playing the piano and harp. She translated a poem written by Michio Mado into English. For the "International Board on Books for Young People (IBBY)" held in India in 1998, although she refrained from attending it due to the fact that a nuclear test was being performed there at the time, she delivered her keynote address, "Memories of Reading Books in My

Childhood," by way of video tape, making a big impression on people all over the world. The keynote address was published as a separate volume, entitled "Hashi wo Kakeru—Kodomo Jidaino Dokusho no Omoide" ("A Bridge of Children's Books —Memories of Reading Books in My Childhood") (Bunshun Bunko in 2009). The Empress has contributed the royalties of this book to the "International Board on Books for Young People," the "Saruhashi Prize—The Association for the Bright Future of Woman Scientists" and the "Round-table Conference for Discussion about Improvement of the Circumstances that Would Allow Female Scientific Researchers to be Active" ("A Bridge of Heart of Empress Michiko" by Mitsuko Watanabe [69].

At the press interview celebrating his 80th birthday, Emperor Akihito stated:

> I, in my position as the Emperor, have been thought sometimes to be lonely, but I was fortunate enough to have found and married my great life partner, who cares about all things precious to me. It has been of great comfort to me that the Empress has always been able to give consideration to my feelings while respecting my position; accordingly, I have been able to fulfill my role as the Emperor.

Everyone says that Emperor Akihito has an amazing memory, just like his father, Emperor Showa. It makes the individual persons who serve and have served the Emperor, including myself, feel deep gratitude that he remembers their works in detail. Makoto Watanabe, who served as a Grand Chamberlain for 10 years from 1996 to 2006, relates the following anecdote in his book [68]:

> Emperor Akihito remembered the surname "Tanabe" of a Japanese-Brazilian poultry farmer whom the Emperor planned to meet when he visited Brazil for the first time in 1978, although the plan was not realized. The Emperor inquired about "Tanabe" when he re-visited Brazil in 1997 for the first time in twenty years. Unfortunately, both Tanabe and his wife had already passed away, but he was able to meet Tanabe's children at Sao Paulo to fulfill his promise.

When Watanabe asked the Emperor Akihito about the time when he had visited the Nasu Imperial Villa for the first time, he answered, "It was the summer of 1937, when I was 3 years old," and talked about his memory of what a great many insects had been living there. In response to Watanabe's inquiry as to these insects, Motofumi Higashisono, a tutor to the Prince, responded that there was a large outbreak of Japanese gold beetles in the Nasu area at that time. It goes without saying that he must have shown his extraordinary memory in regards to the taxonomy of fish, including gobies, to the fullest. This extraordinary memory includes the indispensable ability to remember the places where specimens and materials have been archived and determine whether newly collected specimens are different from old ones. Watanabe, who formerly worked in the foreign service, described the current life and activities of the Imperial Family, just as Sukemasa Irie had previously, because his great-grandfather was the Minister of the Imperial Household and his father was one of Emperor Showa's fellow students in their Gakushuin Primary School days.

Why Did Emperor Akihito Choose to Study Gobies?

I described earlier what made Emperor Showa selectively adopt biology for his research field. But for what reason did Emperor Akihito adopt gobies to be a taxonomical target among a vast variety of fish? He himself stated,

A lot of gobies lived on the beach of Hayama, where the Hayama Imperial Villa was situated, so I became interested in this kind of fish and have continued observation of them,

at a press interview held in 1984, when he was the Crown Prince. According to an article in the newspaper *The Saitama Shinbun*, dated January 1st, 2007, the suggestions by his father, the Emperor Showa, and his former teacher led him to his research on gobies. As for the former episode, at a press interview held in 1986, recollecting his memories of his childhood days, in which he had such a joyful time together with the Emperor Showa on Hayama Beach, he said, "The teaching by my father, Emperor Showa, at that time built up the basics for my knowledge of marine organisms."

Regarding the latter episode, Masayoshi Hayashi, the director of the Yokosuka City Museum, (now moved to "the Biological Laboratory") who who is acquainted with Emperor Akihito in connection with their researches on gobies and has published a number of books, including "The Guidebook to Gobies," stated:

His Majesty told me that research on gobies was suggested by the late Ichiro Tomiyama, who was a professor at the University of Tokyo and a teacher of both Emperor Showa and Emperor Akihito.

According to some sources, Tomiyama suggested to the Emperor Akihito that he select this kind of fish as his research target, giving two reasons; (1) specimens of a wide variety of gobies were capable of being acquired from countries all around the world and (2) gobies might also be collected on Hayama Beach, though I did not hear this directly from Tomiyama.

As described earlier, Tomiyama, with whom the Emperor Showa consulted for many years about his research on hydrozoans at the Biological Laboratory of the Imperial Household, was an expert on the taxonomy of gobies and received his doctoral degree by virtue of a paper entitled "Gobiidae of Japan." The Emperor Akihito once said:

The paper "Gobiidae of Japan" was very useful for me in examining the species of gobies and re-examining the determined species.

At the party held after the International Prize for Biology award ceremony, the Emperor told me, "I was guided by two teachers, Prof. Tomiyama and Dr. Tokiharu Abe, in writing my papers for many years." The Emperor, who has always attempted to break down the barriers separating the Imperial Family from the people, often speaks in the casual manner of the masses, together with the Empress, and addresses members of the general public as Mr., Miss, Mrs., and so on.

Abe, also one of Shigeho Tanaka's disciples, was a taxonomist who was Tomiyama's junior by five years in the Department of Zoology, University of

Tokyo. He worked at the Fishery Experiment Station (the current National Research Institute of Fishery Science) under the Ministry of Agriculture (the current re-organized Ministry of Agriculture, Forestry and Fisheries), and then served as the director of "the Museum of Osakana Fukyu (Fish Information) Center." He continued to visit the Tsukiji Fish market in Tokyo over the period from the time directly after the end of the war to more than 50 years later, and recorded a variety of new species of fish, including globefish. After obtaining specimens of "new species of fish" from a variety of foreign countries, he gave a Japanese name to each of them so as to introduce them internally. It is said that these new fish species numbered more than 200 in total. In addition to Abe, Yasuo Suehiro, called "Dr. Fish" (meaning that he was quite familiar with fish), and Takashi Hibiya, who aimed to achieve full life-cycle eel cultivation at that time, gave lectures to the Emperor. Both of them were professors of Fishery Science, Faculty of Agriculture, University of Tokyo.

Tomiyama asked Kazunori Takagi, a researcher on gobioid fishes, who was then a young lecturer (later promoted to professor) at the Tokyo University of Fisheries (the current Tokyo University of Marine Science and Technology), to deliver lectures to the Emperor (the then-Crown Prince) in his place for three months in 1961, during which he would be unable to be present at the Imperial Palace to deliver lectures himself. Takagi had graduated from the Fisheries Training Center, Agriculture and Trade Ministry, the predecessor of the Tokyo University of Fisheries. He had made a close observation of the sensory system in the goby head and contributed to dramatic progress in the phylogenetic taxonomy of gobies. He also made great strides in the study of gobies in its distribution and ecology. He delivered 7 intensive face-to-face lectures to Crown Prince Akihito on his own thesis "Study on the comparative morphology, phylogenetics, taxonomy, distribution and ecology of fish belonging to the suborder Gobioidei in Japanese waters" that he was preparing at that time. The Crown Prince listened attentively to his lectures from beginning to end. Given that Takagi later wrote, "Perhaps it was a manifestation of his enthusiasm that the Crown Prince wished to use the lectures as a reference in conducting his first research" (quoted from the article "The Emperor's Ichthyological Researches," in a catalog from the Tokyo University of Marine Science and Technology Library [64]), it would seem that the Prince had already commenced his morphological research on gobioid fishes. It is almost certain that these lectures had a great effect on the series of taxonomic and phylogenetic researches on gobies that he conducted later. The Emperor expressed his gratitude to the four supervisors, Tomiyama, Abe, Hibiya and Takagi, for their heart-warming support of his first paper.

Specifically, it might be one of reasons that Ichiro Tomiyama first suggested that the Emperor take the course on gobies (the suborder Gobioidei), which account for the majority of the order Perciformes; accordingly, they are so rich in species and diversity that many of them have not yet been identified. The goby most commonly known in Japan is probably the yellowfin goby (*Acanthogobius flavimanus*), which is widely distributed over a broad area, from Hokkaido to Kyushu, and may excite people of all ages in regard to fishing and their use in delicious dishes, including

Tempura. The mudskipper Mutsugoro (*Boleophthalmus pectinirostris*), which live in mudflats such as the Ariake Sea, are also commonly familiar to us, largely through the books written by Masanori Hata (which have earned him the nickname "Mutsugoro") and many published images. Gobies live in a variety of habitats, including seas, the mouths of rivers and mountain streams. Moreover, although most of them live in the vicinity of benthic regions, they are rich in form and mode of life; for example, some of them, including the whitegirdled goby (*Pterogobius zonoleucus*),swarm away from the sea bottom, mudskippers (*Periophthalmus cantonensis*) jump around mudflats, and *Sicyopterus japonicas*, which live in fast-moving streams, move by means of inchworm motions, sticking to rocks using their mouths and suckers formed out of their pelvic fins. Most gobies are conservatively colored, while some, including the serpentine goby (*Pterogobius elapoides*) and the fire goby (*Nemateleotris magnifica*), are decollated with beautiful colors, much to the delight of the eyes of those who see them in aquariums. They vary a great deal in body length, from about 50 cm for *Acanthogobius hasta* to about 8 mm for a species of candycane pygmy goby (*Trimma cana*, a species from the Indian Ocean), which is the smallest vertebrate animal.

Only 131 species of goby from Japan were included in "Fish Morphology and Hierarchy," published in 1955 by Kiyomatsu Matsubara of the Department of Fisheries, Faculty of Agriculture, Kyoto University, before the Emperor started his research. In contrast, a considerable number of gobies from Japan, specifically, as many as 537 species, were included in "Fish of Japan with Pictorial Keys to the Species, Third Edition," published in 2013. In fact, it is said that the number of recorded gobies may amount to about 600 species. The book was edited by Tetsuji Nakabo of the Kyoto University Museum, and the Emperor also contributed as one of the authors. The number of species in gobies identified throughout the world is somewhere between 2,200 and 4,000.

Gobies have rarely been found as fossils, having speciated in a comparatively later geological age. For this reason, an evolutional and systematical interest, namely, "Why have such a wide variety of species of gobies propagated?", drew increasing attention, and there was great expectation for the discovery of new species. At that time, however, a limited number of researchers throughout the world were studying gobies, and many unidentified specimens were left on the racks of several world museums. Crown Prince Akihito intentionally issued a challenge on the subject, on which others had not being willing to focus, bearing fruit in the form of brilliant achievements in the biological field. The Emperor himself said, "Thinking back now, I selected a very suitable subject," at a press interview held in 1984.

The Emperor described 6 new species of gobies from Japan, *Myersina nigrivirgata, Glossogobius aureus, Pandaka trimaculata, Cristatogobius aurimaculatus, Astrabe flavimaculata* and *Astrabe fasciata*, and almost 20 newly recorded species and newly Japanese-named species from Japan. Moreover, studies on 2 new species from foreign countries were published. It is said that, among the newly Japanese-named species, Akebonohaze (*Nemateleotris decora*) and Gingahaze (*Cryptocentrus cinctus*) were named by Empress Michiko, and Shimaorihaze (*Vanderhorstia ambanoro*) was named by Princess Sayako.

The researches on gobies include: ecological research, which involves field observation; physiological and biochemical research, in which various kinds of instruments and chemicals are used; and newly-emerged molecular biological research. Makoto Watanabe gave an answer to the question, "Why did the Emperor select taxonomy and phylogeny focusing on a comparison between species?", as follows:

> The Emperor told me that the reason why he selected morphology, in which specimens are used, instead of ecology, in which living fish are observed, was that he was better able to control the time spent concentrating on his research in regard to taking time out from his very busy official duties.

As can be obviously seen from what the Emperor said, Emperor Akihito thought about his research in the same way as Emperor Showa. Moreover, Watanabe added:

"Occasionally, we have to talk with the Emperor while he is engaging in research. In such cases, we would hope that we could visit the laboratory to talk with him there without disturbing his research. However, the Emperor always discontinued his research, turned off the light, going up the stairs to his study on the second floor, where he would listen to what we had to say, and then return to the laboratory. Through this practice, he drew a clear line between his private and public duties. The Emperor has never changed this behavior" (quoted from the book written by Makoto Watanabe [68]). The fact that the Emperor would place the utmost importance on his official duties is indicative of his nature. We ordinary persons can't generally easily discontinue our research when we are so thoroughly absorbed in it, even when something preferential comes along. Usually, in such a case, we would request that any interruption be postponed until we reach a stage at which we can take a rest from our work.

The Laboratory of Ichthyology at the Crown Prince's Palace and the Present Biological Laboratory

Similar to Emperor Showa, Emperor Akihito started his research on the taxonomy of gobies in earnest in a room at the Crown Prince's Palace, Akasaka, while he was still Crown Prince, as mentioned earlier. It is said that, in 1961, Kazunori Takagi, as described above, delivered his lecture to the Crown Prince at a modest laboratory situated on the site of an air-raid shelter in the Akasaka Detached Palace. To describe a new species of fish in a paper, it is necessary to allocate a catalog number to it. The room had proved to be the right place for such work, and it had come to be called the "Laboratory of Ichthyology, the Crown Prince's Palace" among colleagues. Subsequently, the laboratory was enlarged so that it occupied 2 inner rooms on the second floor of the Palace; one was used for fish dissection and microscopy and the other was a specimen room, housing many fish tanks and the world's leading specimens of gobies. According to some sources, fish tanks with gobies in them were kept everywhere, including chambers and salons.

Emperor Showa was supported academically by Hirotaro Hattori, chamberlain Tsuchiya, and other specialists in biology, while he was the Crown Prince. Similarly, in addition to the professors described earlier, Teiji Yagi, who graduated from the Department of Fisheries, University of Tokyo, and then did research on the teeth of whales at the Tokyo Medical and Dental University as a research associate, served the Crown Prince as a chamberlain starting in 1964. Later, Yagi served Emperor Akihito, after he was enthroned, as the deputy grand chamberlain. In 1970, he was succeeded by Akira Fuji, who graduated from the Tokyo University of Fisheries and the Postgraduate Course of the Department of Fisheries, University of Tokyo, and also worked as a research associate in the same department. Then, in 1986, Katsusuke Meguro, who graduated from the Department of Fisheries, Nihon University, and was also interested in fishes, served the Crown Prince as a chamberlain. Fuji was a person with wide interests, as seen in the fact that he served as the deputy director of the Japan Society for the Study of the Viola, while I myself have always associated with him within the context of the Misaki Marine Biological Station and meetings in foreign countries. He attended the Crown Prince and Prince Akishino as a councilor while they studied at the University of Oxford. The names of Yagi and Fiji often appear in the "Acknowledgements" of the papers published by Emperor Akihito when he was the Crown Prince. Meguro served the Crown Prince as his technical officer for 20 years, starting from 1966, and was also co-author of many of the papers on gobies published by the Emperor from his Crown Prince days, as will be described later. He would subsequently serve as a chamberlain for the Crown Prince, and then for the Emperor. Kintomo Saionji, who served the Crown Prince as a chamberlain for a period after 1987, and then the Emperor as well, was a great-grandchild of Kinmochi Saionji; in other words, he was a grandchild of Hachiro Saionji, who supported the Emperor Showa in collecting marine organisms and other works, though he was not a specialist in ichthyology. It seems to be a tradition established in the Imperial Court that, when the Imperial Throne is passed on to the Crown Prince, the chamberlains serving the previous Emperor would step down from the position, following the proverb that originated from China: "A loyal subject would not serve another master," with the Crown Prince's chamberlains then succeeding them. When the Imperial Throne was passed on to Crown Prince Akihito and the Imperial era's name was changed from Showa to Heisei in 1989, the grand chamberlain Satoru Yamamoto, who was the second grand chamberlain after Sukemasa Irie, continued to work in that position until his successor could be phased in during a one-year overlapping period.

In line with the Emperor's movement to the Fukiage Imperial Palace from the Crown Prince's Palace in 1993, his laboratory was relocated as well. According to the book written by the former grand chamberlain Watanabe, [68] the Emperor's new laboratory was of middle size, less than 40 m^2, and was built on the first floor of the Fukiage Imperial Palace. A centrally-placed table housed all of the microscopes; specimens were taken out as needed, but were otherwise stored. Fish tanks full of gobies were installed by the window, and the book cabinet was filled with specialized books on fish. Tanks were also placed in other places, such as the dining room, where the Emperor crossbred and reared gobies. He himself changed the water in the tanks, and it is said that he could be seen moving back and forth between the tanks and the water supply carrying a small bucket during rest periods, including the lunch break.

In 1991, after accession to the throne, he inspected every nook and cranny of the Biological Laboratory of the Imperial Household with the Prince Hitachi and the Prince Akishino. It seems that they had a plan to relocate the Biological Laboratory so as to make it available to the public, leaving the specimen room untouched, and to build an Ichthyological Research Institute on the site of the demolished laboratory. Emperor Showa's laboratory was reproduced, with his historical goods put on display in the Emperor Showa Memorial Museum in Tachikawa, Tokyo, open to the public. However, its building, which was not reconstructed, still retains its old appearance. His numerous specimens were moved to several laboratories and museums, such as the "Showa Memorial Institute" at the National Museum of Nature and Science in Tsukuba, the Memorial Museum in Tachikawa and the Yamashina Institute for Ornithology, as described earlier, thanks to a great deal of effort by his chamberlain Ryogo Urabe, as well as Bungo Kawamura and Tatsuya Shimizu, both of whom worked at the Laboratory. His books related to biology were also moved to the National Museum of Nature and Science and other museums after the demise of the Empress Kojun.

A big panel made from a Taiwanese camphor tree still stands near the entrance of the Laboratory, but all of the decorative works, including the deer horns that were previously on the wall of the hall, were removed. Some of the individual rooms of the Laboratory have been assigned to specialists as their research rooms and others are used to house instruments, including an X-ray radiographic apparatus for imaging fish bones, or to rear fish. In the first research room, namely, the main Laboratory (the same one used by Emperor Showa), which is now kept by Masahiro Aizawa, the chief specialist, a large table and microscopes have been installed; this is where the Emperor Akihito, who usually uses the laboratory in the Imperial Palace, engages in his research when he is there, and also where he chats with the staff. In a space that has been formed into a bay window facing the farm field on its south side, fish tanks and a scanning electron microscope, used by him in his Crown Prince days, have been installed. However, it may have been for naught, because the Crown Prince traveled all the way to the National Museum of Nature and Science, bringing his own lunch prepared by Princess Michiko, to use the scanning electron microscope installed there (quoted from "Imperial Family in the Heisei Period" [3]). Concrete specimen rooms on the first and second floors are almost completely filled with goby specimens.

The current staff of the Biological Laboratory consists of three members: Aizawa, a chief specialist, Yuji Ikeda and Takako Sako, both of whom are also specialists. Aizawa, a specialist in the science of fisheries, is a disciple of Yoshiaki Tominaga, who learned directly from Tomiyama, and an expert in the collection and registration of new species, for example, finding a wide variety of new species of *Luciogobius*. In 2007, he changed his career from working at the Coastal Branch of the Natural History Museum and Institute, Chiba, to working at the Laboratory. However, he has recently retired. Yuji Ikeda, a specialist in gobies just like Aizawa, has continuously worked at the Laboratory, still situated at Akasaka, since 1986. Sako, the third specialist, transferred from the Ministry of the Environment and supported the Emperor in his research on the modes of life of animals, including a raccoon dog who lived in the Imperial Palace (recently returned to the previous Ministry). Before Aizawa, Katsuichi Sakamoto was appointed and served as the chief specialist at the Laboratory around 1970. In the beginning of the Heisei Period (around 1990), Akihisa Iwata also worked at the Laboratory; he currently does research on gobies at Kyoto University. The names of these staff members are listed as co-authors in the Emperor's papers and books described later.

The Genealogy of Japanese Ichthyology and the Tokyo University of Fisheries

Now, let's roughly trace the historical path of ichthyological taxonomy, in which the Emperor specializes, in Japan by reference to literature, including "Nihon Dobutsugaku Shi" ("A Japanese History of Zoology"), written by Masuzo Ueno [67].

Similar to other academic fields, knowledge of fisheries had come from China through books on herbalism since about the 6th century. Needless to say, herbalism is a discipline that focuses mainly on medicinal herbs, while it has also provided a foundation for Natural History, including knowledge of animals such as the fish that have developed in the Orient. It is natural that, in Japan, which is surrounded by seas in every direction and is rich in marine resources, individual fish named by fishermen would come to be called by different names from region to region. The names of more than 10 species of fishes, including ayu (sweetfish, *Plecoglosus altivelis*), appear in "Man'yoshu" ("An Anthology of Myriad Leaves"). The Heian Period in the 10th century saw publication of both "Honzowamyo" (the oldest Japanese dictionary of medicine) and "Wamyo Ruijusho" (a Japanese encyclopedia), which resulted in a correspondence between Chinese names and Japanese names through some degree of assimilation of Chinese herbalism into the Japanese style. There was, however, an essential difference in species of fish between China, where the term "fish" represents freshwater fish, and Japan, where it indicates mainly marine fish.

As time passed into the Edo Period, "Honzou Koumoku" ("A Compendium of Medical Material"), written during the Ming Dynasty, was re-introduced and became widespread through the efforts of Ieyasu Tokugawa (the first shogun of the Edo Period), resulting in it regaining in popularity. Meanwhile, during the period of national isolation, the knowledge of natural history developed in the West was introduced into Japan through the Netherlands. Yoshimune Tokugawa, the 8th shogun, was fond of herbalism, and ordered Shohaku Niwa, a medical officer, to conduct a survey on the products from various areas of Japan. Some of the results of the surveys conducted individually in the feudal domains were compiled for storage and are still available. Among them, 110 species of fish were recorded at Nagato and 119 species at Suo, Choshu Domain (the current Yamaguchi Prefecture). Linné published his famous book "Systema Naturae" while Yoshimune was alive (1735).

During the period from the end of the 17th century to the beginning of the 19th century, animals, including fish and plants, living in Japan were introduced into the Western world by doctors, who came to the Dutch Trading House situated at Dejima, Nagasaki, which was exclusively open to foreign countries at that time. Engelbert Kaempfer, a German doctor, who came to Japan in 1690, made two round trips between Nagasaki and Edo (the current Tokyo), accompanying the kapitan (chief) of the Dutch Trading House, and collected and observed animals and plants along the way as much as possible. Over 45 species of fish appear in "The History of Japan," which he wrote after returning home. The book is enhanced by illustrations from "Kinmo Zui," drawn by Tekisai Nakamura. "Kinmo Zui" is an illustrated reference book of a wide variety of organisms, ranging from medicinal herbs, foods and drinks to familiar animals.

Carl Peter Thunberg, who was a Swedish disciple of Linné and who came to Japan 85 years later, described several dozen species of fish, in addition to plants, his specialty, in his book "Fauna Japonica." The plants from Japan that were given species names by Linné were sent to him by Thunberg. At that time, the restrictions imposed by the Edo Shogunate had been somewhat relaxed, and he was able to directly observe animals and plants from Japan and have deep exchanges with Japanese scholars.

Philip Franz von Siebold, another German doctor, who came to the Dutch Trading House in 1823, had exchanges with many Japanese people and greatly inspired them. Moreover, he brought a great number of animal and plant specimens to Leiden University, the Netherlands. The specimens of fish that he brought amounted to 358 species, and Coenraad Jacob Temminck and Herman Schlegel worked on the section on fish in "Fauna Japonica," which was edited by Siebold after his return home. Tanshu Kurimoto, a medical officer in Edo who edited "Gyofu" ("Records of Fishes"), which featured many noteworthy sketches of fish, showed Siebold his book upon meeting him. The book was never published, with the exception of a small portion. As a result, it has remained unknown to the public, an extremely regrettable fact.

As the Edo Period changed over to the Meiji Period, Japan made efforts to absorb new knowledge from Western countries, while herbalism gradually faded into obscurity as a discipline of the old world, to which a high value had been attached in the past. Under such national conditions, some of the people playing important roles in the society were so-called hired foreign teachers. As described earlier, German teachers, including Franz Martin Hilgendorf and Ludwig Döderlein, who came to the Tokyo Medical School (the predecessor of the Medical School, University of Tokyo), in 1873 and 1879, respectively, delivered lectures on modern zoology and botany to the students, and collected marine animals, not only in Sagami Bay but also in fish markets, where they viewed the specimens on display with enthusiasm. In a book published after they returned home, about 40 new species of fish in total are detailed (Cf. Fauna Sagamiana [62]).

After mastering German, Shin'nosuke Matsubara (1853–1916), from the Matsue Domain (the current Shimane), entered the Tokyo Medical School and learned zoology, especially ichthyological taxonomy, while acting as an interpreter for the lectures delivered by Hilgendorf. Later, he studied the science of fisheries in Germany and, after returning home, was put in charge of ichthyology and fish-raising at Suisan Denshujyo (The Training Center for Fisheries), the Ministry of Agriculture and Commerce. Matsubara and Hilgendorf co-authored the "Catalog of Fishes of Japan" in German used at the International Fisheries Exposition held in Berlin at that time. It was the first catalog in which Latin scientific names were given to fish by a Japanese scientist. Moreover, he was committed to the foundation of the Imperial Fisheries Institute, the predecessor of the Tokyo University of Fisheries (the present Tokyo University of Marine Science and Technology) and, in 1903, was appointed as the first director of the Institute. Thus, a foundation for ichthyology was built up at the Tokyo University of Fisheries.

Meanwhile, Morse and Whitman, the two American teachers, played active roles in the Department of Zoology, Faculty of Science, University of Tokyo. There was a complicated factor behind the appointment of Morse as the first professor at the Department of Zoology. In fact, Ryokichi Yatabe, the first professor at the Department of Botany, had recommended David Starr Jordan (1851–1931), an ichthyologist, who was his fellow alumnus from Cornell University, but Morse was appointed while Jordan was still indecisive about accepting the position. On the other hand, Jordan was later appointed as the president of Stanford University while he was still young and highly regarded as an outstanding figure in the field of ichthyological taxonomy. He visited Japan in 1900, with John Snyder, his disciple, to collect fishes and actively toured the Japanese Islands from Hokkaido to Kyushu. He also visited the Misaki Marine Biological Station. He collected about 1,000 species of fishes in Japan, among which more than 200 were new species. The results of his collection were published in his many papers and compiled into an English monograph, "A Catalogue of the Fishes of Japan," in 1913. He visited Japan again in 1922.

One of the co-authors of the afore-mentioned book was Shigeho Tanaka (1878–1974), who was known as the father of ichthyological taxonomy in Japan (Fig. 2.8). It is said that, being from Kochi, he contended with Torahiko Terada, one of his classmates, for the top seat in their junior high school days. Terada was a physicist and a disciple of Soseki Natsume. Tanaka, after graduating from the Department of Zoology, University of Tokyo, started his taxonomical research on fish in 1903. When he visited the US and commissioned Jordan to proofread the manuscript of his English monograph, Jordan requested that he add his name to the list of co-authors. Tanaka accepted this request, and the book was published with Jordan's name added. In fact, Jordan seemed only to have refined a very few points. Tanaka collected a vast quantity of specimens of fish at the University of Tokyo and described about 170 species as new species. Moreover, he educated his disciples, who later supported the Emperor Akihito in his research on taxonomy of gobies, described later. He also identified the fish collected by Emperor Showa. Tanaka continued to publish "Figures and Descriptions of the Fishes of Japan, including the

Ryukyu Islands, Bonin Islands, Formosa, Kurile Islands, Korea, and Southern Sakhalin," comprised of 48 volumes, in which about 300 species of fish were described in detail, one by one, during the period from 1911 to 1931. In 1938, he was appointed as the director of the Misaki Marine Biological Station.

In addition, in 1900, when Jordan visited Japan, Bashford Dean, an ichthyologist who specialized in sharks and whose father was a friend of Morse's, also visited the Misaki Marine Biological Station. He was wealthy enough and generous enough to donate a collection ship, the "Arai Maru," to the station. The house that he donated when visiting Japan again 5 years later is commonly known as the "Dean's Residence," and I myself lived there at one time in my young days.

The top disciple of Shigeho Tanaka was, perhaps, Toshiji Kamohara, who was also from Kochi and was Tanaka's junior at the university. Since 1928, when he was appointed as a professor at Kochi Higher School (under the old system, the predecessor of the current Kochi University), following a brief interval after graduation, Kamohara began to classify fishes under the guidance of Tanaka. He published books such as "Colored Illustrations of the Fishes of Japan" and "Colored Illustrations of the Marine Fishes of Japan" and cultivated a tradition of ichthyology at Kochi University. An enormous number of specimens of collected fishes are stored at the university, forming a combined fish collection with those at the University of Tokyo and others in Japan.

The next generation disciples were Ichiro Tomiyama, who we have already described several times before, and Tokiharu Abe, both of whom had also studied under Tanaka in their graduation researches starting in 1929 and 1934, respectively. They took over Tanaka's work on "Figures and Descriptions of the Fishes of Japan, including the Ryukyu Islands, Bonin Islands, Formosa, Kurile Islands, Korea, and Southern Sakhalin," and published up to Volume 59 (1958). Tomiyama classified a vast amount of fish specimens collected by Tanaka, preserved in formalin, and moved them to another specimen room in the current University Museum. A description of these efforts appears in a monthly publication "UP" from that time, published by the University of Tokyo Press [66].

Juro Ishida, one of my mentors, specialized in developmental biochemistry. However, in his young days, he brought many specimens of Salmonidae, including *Salvelinus malma*, from Karafuto (now Sakhalin) back to the University of Tokyo and conducted his research on them under the guidance of Tanaka. Looking back, when the subject "Examine the life modes of sticklebacks" was given at a seminar a short time after I entered the University in 1950, I asked Tokiharu Abe to show me the specimens of three-spined stickleback (*Gastrosteus aculeatus*), amur stickleback (*Pangitius sineesis*) and hariyo (*Gasterosteus microcephalus*) among the numerous fish specimens stored in the Department of Zoology at that time. With Abe's permission, I was able to observe them. Now, many habitats of these fishes have been lost here and there.

Tomiyama thought that it was really unfortunate that, among Tanaka's other disciples, Shigeo Masuda, who had graduated from the University around the start of the Pacific War, died young on an unknown front during the war. Masuda had been interested in fish since his childhood and wrote many excellent papers on groupers. Thus, no successor to Abe was cultivated in the Department of Zoology, University of Tokyo, until Yoshiaki Tominaga joined Tomiyama's laboratory as a graduate student in 1959. Tominaga, who specialized in sweepers (Pempheridae), which are difficult to classify, took the action of leading a group of people who helped him to move the specimens from the Department of Zoology to the newly-built University Museum, University of Tokyo. Unfortunately, he also died young. As already described, Masahiro Aizawa, the contemporary (now former) chief specialist at the Biological Laboratory, had studied under Tominaga. Two years later, Ryoichi Arai, who would eventually be appointed as the director of the Department of Zoological Research, National Museum of Nature and Science, moved to the Graduate School from the Tokyo University of Fisheries, and then, 8 years after that, Torao Sato carried out research on ichthyological taxonomy, mainly at the Misaki Marine Biological Station, under the guidance of Abe. However, it seems that no subsequent researchers have followed the path of ichthyological taxonomy at the Department of Zoology, though it should be noted that many experimental researches using fish, especially Japanese medaka (*Oryzias latipes*), are being conducted. The specimens of fish displayed in the University Museum, University of Tokyo, have been maintained by Abe, Tominaga, Sato, et al., and recently, Kazuo Sakamoto, the director of the Fish Information Center Museum, and others have been put in charge of managing the specimens. Additionally, Ryoichi Arai, who, after retirement, moved to the Division of Evolutionary Biology, University of Tokyo, from the National Museum of Nature and Science, also continues to conduct his research on fish at the University Museum. The Emperor also visited the Museum when Abe was alive. The names of Abe, Tominaga, Arai, et al., often appear in the papers written by the Emperor in his Crown Prince days.

How was taxonomy positioned in the Imperial Fisheries Institute, at which Shin'nosuke Matsubara worked as the first president, and in the subsequent Tokyo University of Fisheries? Kanzo Uchimura, who is known as a Christian missionary and delivered his lectures with Matsubara at Suisan Denshujyo, had already compiled "Catalogue of Japanese Fishes," which shows the correspondence between the scientific names and Japanese names of ca. 640 species of fish, but it had not been published at that time. After a lapse of years, Kiyomatsu Matsubara, who graduated from the Training Center for Fisheries in 1929 and worked at the Imperial Fisheries Institute as a research associate for Arata Terao who gave the name *"Sympasiphaea imperialis"* to the prawn discovered by Emperor Showa (Fig. 2.9). In 1943, he co-authored "Keys to the Fishes and Fish-like Animals of Japan" with Yaichiro Okada, one of his seniors and a professor at the Tokyo Higher Normal School (later the Tokyo University of Education, and then the University of Tsukuba). Four years later, he was appointed as the first professor at the Laboratory of Aquatic Biology, Department of Fisheries, Kyoto University, and delivered his lectures on systematic

Fig. 2.9 Kiyomtatsu Matsubara (Courtesy of Tetsuji Nakabo)

taxonomy, inspiring many disciples. Meanwhile, he wrote "Fish Morphology and Hierarchy," which is a must-read book for taxonomists, and "Gyoruigaku" ("Ichthyology"), a book in two parts.

Thanks to his efforts, many fish specimens, especially demersal fishes in the sea area around Japan, and literature related to them were collected at Kyoto University and, after the war, the most favorable environment for ichthyologic taxonomy was established in Japan. For example, Akira Ochiai, who studied flatfish (Platycephalidae) at Kyoto University, went on, after graduation, to play an important role at Kochi University.

Tamiji Kawamura, a freshwater biologist and the first professor at the Department of Zoology, Kyoto University, had already collected various kinds of fishes, including freshwater fishes. The specimens of these fishes are now stored at the Maizuru Fisheries Research Station of Kyoto University and the Kyoto University Museum, and Tetsuji Nakabo, who followed a tradition of ichthyology under the guidance of Tamotsu Iwai, a successor to Matsubara, serves as a professor at the University Museum. He compiled "Fishes of Japan with Pictorial Keys to the Species" in 1993, the third edition of which has been published in 2013. The Emperor wrote the section on gobies with his colleagues at the Biological Laboratory. Incidentally, Nakabo now serves the Emperor as an advisor in his research on fishes.

At the Tokyo University of Fisheries, after Matsubara left, Itsuo Kubo, who later delivered lectures on zoology and fishery resource science, succeeded him as a research associate. When he was appointed as a professor at the Laboratory of Fishery Resource Science, Kazunori Takagi served as a research associate, as described earlier. Takagi edited the "Dictionary of Japanese Fish Names and Their Foreign Equivalents," published by the Ichthyological Society of Japan in 1981, with Fujio Yasuda, who graduated from the Tokyo University of Fisheries alongside Takagi, Tominaga, a disciple of Tomiyama, and Teruya Uyeno, who served as the director of the Department of Paleontology, the National Museum of Nature and Science. The Zoological Laboratory was re-organized into the Laboratory of Ichthyology in the mid-1960s and, after the Tokyo University of Fisheries was renamed as the Tokyo University of Marine Science and Technology, was further re-organized into the Laboratory of Ichthyology, Division of Marine Biology, Department of Marine Environment, Faculty of Marine Science. Nowadays, not only the names of universities, but also those of their laboratories, are too long to easily identify. I hope that they will someday be shortened for easy identification.

The successors of Matsubara at the Laboratory of Ichthyology include Reizo Ishiyama, who classified skates of the family Rajidae, and Yasuhiko Taki, who classified freshwater fish of Southeast Asia, did research on biogeography, gave guidance about research on fish to the Prince Akishino and provided the Emperor with specimens of gobies. Hiroshi Kono, who conducted researches on the systematic taxonomy of mackerels of the family Scombridae and then on the morphology and ecology of fry, said:

"Our duty is to develop 'interdisciplinary ichthyology' from the standpoint of a 'wider perspective', which Professor Shin'nosuke Matsubara, the first director of the Imperial Fisheries Institute, has taught by setting a personal example, with the deeply held enthusiasm for ichthyological taxonomy that was passed down to us by Kiyomatsu Matsubara" [64]. The Tokyo University of Marine Science and Technology held a special exhibition entitled "The Emperor's Ichthyological Researches" in 2009.

In contrast, it seemed that, at the Sapporo Agricultural College, a predecessor of Hokkaido University, fishery was initially considered to be the domain of fishermen, and, accordingly, the science of fisheries was also viewed as unimportant. Against the background described above, John Cutter, one of the hired foreign teachers, who came to Japan and arrived at his post at the college in 1878, only assigned half as many of his zoological lectures to the science of fisheries as would have been expected. Kanzo Uchimura, described earlier, was a graduate of the Sapporo Agricultural College in the early days. The Department of Fisheries was created in 1907, and Toyoji Hikita made a great contribution to research on fish living in the northern waters. The tradition of the science of fisheries at the Department was succeeded by Kunio Amaoka, who graduated from Kyoto University. He collected specimens equivalent in size to the collections stored at the University of Tokyo and Kyoto University, and built one of the core institutes focusing on systematic ichthyology at Hokkaido University. His disciples are also playing important roles at the institutes of Kochi University, Kagoshima University, etc. Keiichi Matsuura, who engaged in classifying fish at the National Museum of Nature and Science, also graduated from the Department of Fisheries, Hokkaido University.

In Japan, the methodology of ichthyology has changed from a focus on close examination of fish caught in the sea around Japan to a strong commitment to investigate specimens of fishes caught in various fishing grounds throughout the world; the opportunity to do so sprang from the fact that the fishing boats in Japan were enlarged around 1960, allowing fishermen to sail to seas around the world (quoted from the article by Akira Ochiai [50]). In addition, the methodology of systematic taxonomy has also changed from morphological and ecological comparisons between fish to biochemical methods using isozymes and molecular biological methods using DNA sequences.

Emperor Akihito as an Academic Researcher

When Kazunori Takagi, who had graduated from the Tokyo University of Fisheries, was continuing his studies at the Faculty of Fisheries, Kyoto University, situated at Maizuru, while writing his doctoral thesis after delivering his lecture to Crown Prince Akihito at the Crown Prince's Palace, the Crown Prince himself would call on him at around 8 almost every evening and put many questions to him about his research. When the Crown Prince was preparing to submit his paper to academic journals, Takagi visited him with Ichiro Tomiyama, to give him guidances on the rough draft of his manuscript, and the discussion went on from 3 in the afternoon to the middle of the night (quoted from the article by Keiichi Matsuura [37]). This anecdote lays bare the Prince's enthusiasm as a sincere scientist, who pays careful attention to everything.

The Emperor published his first paper, "On the scapula of gobiid fishes," in 1963. (Gobioid corresponds to Suborder and gobiid to Family). This paper was intended to present the fundamental data for classification of fish of the family Gobiidae, focusing on pectoral fins, to which less attention had been paid so far, in order to make comparisons among as many gobies as possible. To write this paper, he boiled the specimens preserved in alcohol so as to easily separate the bones, passed them through the series of formalin and sodium hydroxide, and finally used the staining technique in which the bones are dyed red with Alizarin [red dye extracted from the root of madder (*Rubia tinctrum*)], which had just come into use at that time. It was through such time- and labor-consuming processes that he was able to make his observations of the scapula specimens. In 1967, he also published another subsequent paper on the scapula. Both of them examined 150 species of gobies in detail, one by one. Two years later, he tried to classify gobies based on the bones, including those other than the scapula, using 106 species. A far greater number of individuals than those described above must have been used for the actual observation, indicating that this type of work requires steady effort and patience.

It is said that, in his Crown Prince days, the Emperor was often seen in the early morning wearing working clothes and rubber boots while collecting gobies in the rivers near the Hayama Villa or in rural areas (Fig. 2.10). Anyone who saw him collecting gobies perhaps did not realize that he was the Crown Prince. This is a clear indication that, just like Emperor Showa, he puts great effort into field work, in

Fig. 2.10 Emperor Akihito and others collecting gobies in a scoop net in the vicinity of the mouth of a stream near his accommodation while attending the Kagoshima National Sports Festival in 1972 (Courtesy of the Asahi Shimbun Company)

addition to the work done in the laboratory. He not only collects gobies, but also does research on their biology from a wide variety of viewpoints, by regularly observing the behaviors of those living in the aquariums, through hybridization experiments, observation of developmental processes, etc. He is also keen on acquiring experimental techniques and skills and has used x-ray equipment to examine the skeletons of gobies and a scanning electron microscope to observe the microstructures of their sensory organs.

He himself drew the original sketches of gobies featured in his papers and books. Some of these sketches are shown in Fig. 2.11. All of the light and dark patterns on the fishes are represented by stippling, namely, pointillism. Moreover, he achieves a sense of roundness, a three-dimensional element, in his two-dimensional flat sketches. This work must have required him to be patient to a considerable degree, even though he has been good at drawing since his childhood days. My own experience with sketching during laboratory exercises in my university days was one of extreme difficulty, with very little to show for it in the end. This fact shows us the intensity with which Emperor Akihito has been carrying out his research within the limited time that he can carve out between his official duties.

Fig. 2.11 Illustrations of dusky tripletooth gobies, *Tridentiger obscurus*, drawn by Emperor Akihito (From "Fishes of Japan with Pictorial Keys to the Species 3rd Edition" courtesy of Akihito et al.)

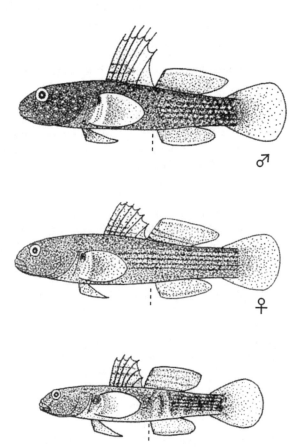

Ryoichi Arai, who worked at the National Museum of Nature and Science, wrote, in 1990:

"His serious attitude about the research on gobies may be well understood by reading his papers from two viewpoints: first, an enormous number of fish were used in his research, and second, based on close observations, individual data are described in detail. With the exception of his papers that are written in English, sentences in the double-negative mode, for example, "I do not necessarily consider that … is not applicable…," are often used in the "Discussion" chapters of his Japanese papers. From the standpoint of scientific papers, this type of sentence is apparently a roundabout expression, and too scrupulous attention to details tends to make it difficult to understand the overall context of his papers. However, the Emperor intentionally uses such expressions to give emphasis to the correct description of scientific knowledge and limitation to the analogy of relationships among gobies based on their morphological features alone" [3]. Later, the Emperor himself addressed the latter issue in a course of works on molecular phylogeny using mitochondrial DNA.

In terms of taxonomy, specimens used to describe a certain species of organism are referred to as "type specimens." One of these designations is as a "holotype (standard) specimen." To identify a specimen discovered later or to describe a new species, it is often required that a close comparison be made with their corresponding holotype specimens. Other individual specimens collected at the same time are referred to as para-type (subsidiary) specimens, which are of value as well.

By the way, as described above, most animals and plants in Japan bear the species names given to them by Western scientists during the Edo through to the Meiji and Taisho Periods. The holotype and para-type specimens of these species are stored in Western museums, including the Natural History Museum, London. For this reason, Japanese taxonomists had to spend much time and labor visiting foreign countries to observe the desired types of specimen or else find a curator who would send the specimens to them in order that they might be able to determine the identity of the targeted organism, namely, to examine which type of specimen would or would not correspond to it. Emperor Akihito, who has experienced the same difficulty in his work classifying gobies, hopes that the type specimens of the animals and plants described in Japan will eventually be strictly stored in suitable storage places in Japan as much as is possible. Unfortunately, in Japan, where no natural history museum has taken solid root, less attention has been paid to specimens, and even type specimens have not been carefully managed in many cases.

As described later, the Emperor co-authored a paper entitled "*Glossogobius sparsipapillus*, a new species of goby from Vietnam," with Katsusuke Meguro, his chamberlain in his Crown Prince days. The paper represents observations of 6 specimens of goby; 5 were collected by Nobuyuki Kawamoto, a fish physiologist, and 1 by Yasuhiko Taki, who was described earlier in the section "Tokyo University of Fisheries," in South Vietnam. Among these specimens, one of the gobies collected by Kawamoto was designated as a holotype specimen and 3 were para-type specimens, and they were all stored in the National Museum of Nature and Science. However, the Crown Prince's intention, "I hope to donate the specimen collected (by Taki) to Vietnam as a para-type specimen. So, make inquiries about the appropriate museum in Vietnam for the donation," was conveyed by the officials of the Crown Prince's Palace. Following his wishes, the specimen was ultimately registered and stored in the Zoological Museum, Vietnam National University, Hanoi, with great effort shortly after the end of the Vietnam War. When Taki asked Meguro, the then-Crown Prince's chamberlain, the reason why the Emperor had expressed such an intention, his answer was as follows:

> Most of the animals and plants of Japan were described as new species by Western scientists in the past and their type specimens are displayed in Western museums. At present, Japanese researchers have to expend considerable time and labor in order to get an opportunity to closely observe the specimens. I would prefer that the Vietnamese researchers not experience the same difficulty as we did (Yasuhiko Taki [59]).

This story gives a true account of his thoughtful consideration as a scientist for researchers in other countries. Later, the Emperor had the chance to see this specimen again when he visited Vietnam.

The Emperor made an oral presentation at the annual meeting of the Ichthyological Society of Japan for the first time in 1985. According to some sources, even though various restrictions are imposed on him, he would like nothing better than to be treated as just another researcher, in the same manner as other members when participating in academic societies. Although I did not have the opportunity to attend the meeting in which the Emperor participated, I was told that he sat on one of the seats for students, together with the other members, and raised his hand when he wanted to ask a question.

A study group on fish classification, called "The Workshop on Fish Taxonomy," which has been managed by Teruya Uyeno and Ryoichi Arai, both of whom worked at the National Museum of Nature and Science, has been held once a month at the Museum for over 30 years. The Emperor attends almost every workshop as one of the members, and once made an hour-plus-long presentation on the taxonomy of gobies. In addition, he used to enter into active discussions with the other scientists, including the postgraduates. According to some sources, "whenever he had to make a presentation to the Workshop, he would do a careful dry run at the Palace, even though it was not as if he was giving a lecture at the academic society" (quoted from the article by Ryoichi Arai [3]). Even in 2013, when he was very busy as usual, he attended the Workshop as many as 6 times.

Among scientists in their 70s and 80s, those, including myself, who continue to conduct their research, are few in number in the field of natural science. It is true, especially in the field of experimental science, that scientists tend to step away from active research when reaching the retiring age of 60 or 70, in many cases mainly because it would be difficult for them to make full use of various scientific instruments. And only a small number of members (infrequently) show up at the annual academic meetings, for which they have registered. Knowing that the Emperor, who is still the busiest person in the country, despite being over 80, is passionate about his research on fish taxonomy and the preservation of organisms, I must sincerely take my hat off to him. Additionally, the ichthyologists are invited to all celebration events on the Emperor's Birthday every year.

Kunimasu, Bluegill, Tilapia and Coelacanth

At the press conference held for the celebration of his birthday at the end of 2010, the Emperor gave his impressions of that year:

> Toward the end of the International Year of Biodiversity, a member of the freshwater fish of Japan was added to the list. It is Kunimasu (*Onchorhynchus nerka kawamurae*), as recently reported by the media. Kunimasu, which had lived only in Lake Tazawa, Akita Prefecture, in the past, became extinct due to the mixing of strong acidic river water into Lake Tazawa to increase the volume of water so as to use the lake for power generation from 1940 until 1948. However, it was confirmed, at the beginning of this year, that the eggs of Kunimasu had been transplanted into Lake Sai, Yamanashi Prefecture, resulting in the species' successful survival there to this day. It is not too much to say that Kunimasu is a miracle fish. I have a memory of Kunimasu from the age of 12, when I was reading "Shonen Kagaku Monogatari" ("Scientific Story for Teenagers"), written by Masamitsu Oshima. This book asserted that the Kuminasu living in Lake Tazawa would be destined for extinction in the future due to the inflow of acidic water. This suggestion has remained deeply in my mind. Sixty five years later, we heard good news about the survival of Kunimasu.

He expressed his respect for the people who had contributed to the determination of whether the fish discovered was Kunimasu, and spoke of the future protective measures for these fish.

Tamiji Kawamura, who worked at the Department of Zoology, Faculty of Science, Kyoto University, presented specimens of 3 male Kunimasu fish from Lake Tazawa to David Starr Jordan when he visited Japan in 1922. In 1925, Jordan and his disciple, Snyder, after observation and examination of them, published a paper stating that Kunimasu was a new species of the family Salmonidae.

The Satake Domain, which had been governing the territory, including Lake Tazawa, had known for a long time that blackish trouts (Kunimasu) lived only in that lake. In recent years, the eggs of Kunimasu were hatched and the harvested fry were released into rivers by the Akita Prefectural Fisheries Experimental Station. The adult fish, about 30 cm in body length, always stay mature, being able to be fertilized all year round and spawn eggs in deep water, unlike other species of salmon and trout. Meanwhile, as the Emperor had foretold, Kunimasu became extinct due to acidified water caused by the works conducted to enhance electric power during the war. However, a record remains that describes that a large number of eyed eggs had been delivered to Lake Motosu and Lake Sai (Yamanashi), as well as Lake Biwa (Shiga) and Lake Nojiri (Nagano). Based on this record, the local Tazawako Town Office offered a 5 million yen reward for anyone who could find Kunimasu, but specimens failed to turn up, and the Town Office finally threw up their hands in defeat. Kunimasu was described in the Red Data Book (of endangered species) as an extinct species.

The 9 specimens of Kunimasu fish, both male and female, are stored at the University Museum, Kyoto University. Tetsuji Nakabo tried to reproduce Kunimasu, at least in images by means of computer graphics (CG), but has failed to achieve his aim. Against this background, in March 2010, 2 black kokanees, *Oncorhynchus nerka* (locally called "black trout"), caught in Lake Sai by chance were brought to Kyoto University. Even though they were considered possibly to be Kunimasu based on the historical record of transplantation, Nakabo could not make any prompt determination, because their size was only 20 cm in body length and some of adult kokanees might also turn black. To this end, he asked the president of the Fishermen's Union of Lake Sai to provide "black kokanees," obtaining 5 fish, a mix of male and female. It was clarified that they spawned their eggs at sites other than shallow waters, indicating that they coincided ecologically with Kunimasu. Moreover, the morphological comparison between both species and a DNA analysis allowed him to confirm that they were of different species from kokanees. They were discovered for the first time in as long as 70 years since the transplantation from Lake Tazawa. As an aside, in Lake Motosu, Kunimasu hybridized with kokanees at an early stage, while in Lake Sai, both species were ecologically separated from each other and were miraculously protected from being hybridized, allowing Kunimasu to survive as an independent species (Tetsuji Nakabo [48]).

In contrast to Kunimasu, one of the fish now designated as a specified foreign organism to be exterminated is the bluegill of the family Centrarchidae, the order Perciformes. The bluegill is flat-shaped and about 20 cm in body length. The name "bluegill" ("blue branchia") was given to it on account of the blue projections on their operculums. They live in stagnant freshwater environments. Fifteen bluegills native to the Mississippi river system were presented to Crown Prince Akihito by the Mayor of Chicago during the Prince's foreign tour in 1960. The Crown Prince brought them back to Japan and donated them to the Freshwater Fisheries Research Laboratory, Fisheries Agency. The laboratory tried to raise them, in hopes of providing a new kind of food for the public, delivering them to local laboratories. However, the bluegill, which develops slowly, turned out not to be suitable for aquaculture. It is said that the hatched fishes were eventually released into Lake Ippeki, Shizuoka, several years later.

Since then, many lovers of fishing have released bluegills into lakes and marshes all over Japan, on the grounds that they would help them attract and catch large-mouth bass, also designated as a specified foreign organism, causing them to be distributed throughout Japan. The recent analysis of mitochondrial DNA revealed that all of the bluegills living in Japan originated from those brought by the Crown Prince. At the sites where they were released, the native fish species have disappeared, and, instead, only bluegills and black basses have remained. To maintain species diversity, including preservation of valuable species native to Japan, these species have to be eliminated to the greatest degree possible.

At the "National Meeting for a Healthy Ocean," held in 2007, the Emperor stated that:

> About 50 years ago, I brought the bluegills into Japan from the US and donated them to the Laboratory of the Fisheries Agency. Initially, they were expected to be edible, and their aquaculture was begun, but I feel distress that things turned out in this way.

The Emperor's disappointment is understandable, given that he pays special attention to fish designated as special, natural national treasures, including *Rhodeus ocellatus kurumeus*.

However, it seems that the reason why *R. ocellatus kurumeus* became extinct is not that they were overcome in the struggle for existence by bluegills, but rather by *R. ocellatus*, which is a non-native fish introduced from China. However, some survivors among the purebred individuals of *R. ocellatus kurumeus* were carefully cultured by the Emperor and the Prince Hitachi in the pond of the Akasaka Detached Palace and have already been reintroduced into their original habitats.

But how about tilapia, of the family Cichlid, the order Perciformes, to which bluegills also belong? Tilapia were introduced into various areas of the world and cultivated there to address the problem of food shortage after World War II. In Japan, Nile tilapias (*Oreochromis niloticus*) were cultivated for food for the citizens. They having a similar appearance to black porgy (*Acanthopagrus schlegeli*), also called Izumidai, are pretty good in terms of taste, and were distributed throughout Japan. However, recently, red sea breams (*Pagrus major*) have been cultivated in large quantities and have come to dominate the market over Nile tilapia. Tilapia are, of course, different from the red sea bream of Japan in that they live in freshwater or brackish water, but not in seawater. Being unable to live in low temperature water, they still survive in Okinawa and hot-spring areas and are used not only for food, but also as an experimental animal species. In some cases, they increase in large numbers out of doors and are treated as one of the blacklisted, specified foreign organisms, because they have the potential to disrupt the ecosystem.

In 1964, the Crown Prince visited Thailand. At that time, the food situation in Thailand was very serious, especially in regard to protein, for tribes living in the depths of mountains, and, to address this situation, some tilapias were brought to the fisheries laboratories there. The Prince presented 50 Nile tilapias raised in the Imperial Palace that he had watched grow with his own eyes to Thailand's King Bhumibol Adulyadej (Rama IX) and encouraged the King to cultivate them, because he thought that they were more suitable for food, being larger than the tilapia usually found in Thailand. At present, in Thailand, these tilapias have propagated everywhere and have become common food items, thus the Prince's contribution is greatly appreciated by the Thai people. Based on this episode, Chinese Thais represent the tilapia as "仁魚" in Kanji, after Emperor Akihito (明仁), and call it "pura-nin" in Thai. Following the precedent, the Japanese Government presented half a million tilapias to the Republic of Bangladesh so as to relieve a food crisis that was occurring in 1973.

The coelacanth is a fish that appeared in the Devonian Period about 400 million years ago, and it was believed that this species of fish became extinct around 55 million years ago. However, in 1938, following a chance appearance by a coelacanth in a fishing net, it was confirmed that the fish had continued to survive on the east coast of South Africa, creating a sensation around the world. Since then, coelacanths have been continuously discovered near the Comoro Islands in the Indian Ocean and in other places, and the survival of another species was ascertained in the sea near Sulawesi, Indonesia, in 1997. These truly are "living fossils." The specimens of coelacanth have been transported to museums and laboratories all over Japan. At the Tokyo Institute of Technology (TIT), Norihiro Okada and his colleagues, under Okada's leadership, asked the Fishery Laboratory of Tanzania to send them frozen specimens of coelacanth, and they are currently making researches on the obtained specimens. In 2009, when the Emperor visited the Suzukake Campus, TIT, Masataka Okabe, a professor at the Anatomical Science Laboratory, Jikei University, School of Medicine, et al., dissected the female individual. The Emperor, who had been interested in coelacanths, attended the dissection wearing a white lab coat and rubber gloves, and asked various kinds of detailed questions. Since then, TIT, the National Institute of Genetics and other institutes and laboratories have, for the first time in the world, decoded the coelacanth's entire genome using the specimen stored in Japan. The result suggests that the speed of its genetic evolution is extremely slow and that the genes involved in limb development and olfaction in land animals have already expressed.

Achievements of Emperor Akihito

Unlike in the case of Emperor Showa, most of Emperor Akihito's achievements have been original papers that he has written, which have been peer reviewed by specialists in their corresponding fields and then published in such academic journals as "The Japanese Journal of Ichthyology", "Ichytological Research" and a world-renowned journal "Gene." Moreover, being the primary author, he has to assume full responsibility for most of these papers. Special treatment has not been given to him at all in deciding whether or not they are accepted for publication. For this reason, it has sometimes taken a long time to refine them in the course of exchanging questions and answers with the reviewers until the papers are finally accepted and printed. Takashi Gojyobori, one of the co-authors of his papers, who has been engaged in joint research with the Emperor for about 20 years, stated:

> The Emperor, who considers that it is natural for scientific papers to undergo strict review, refines his work through repeated discussions with us whenever the manuscripts are sent back after being stringently edited, and then submits the final versions (quoted from Our Imperial Family, No. 45 [53]).

The first original paper that he wrote by himself, which was published under the single-author name Prince Akihito, was
"On the scapula of gobiid fishes," Japanese Journal of Ichthyology, Vol. 11, pp 1–26, 1963. https://doi.org/10.11369/jji1950.11.1

His subsequent single-author papers that have been published are listed chronologically below:

"*Glossogobius biocellatus* (Cuvier et Valenciennes) collected in Kyushu," Japanese Journal of Ichthyology, Vol. 12, pp 1–6, 1964. https://doi.org/10.11369/jji1950.12.1
"On the scientific name of a gobiid fish named "urohaze,"", Japanese Journal of Ichthyology, Vol. 13, pp 73–101, 1966. https://doi.org/10.11369/jji1950.13.73
"On four species of the gobiid fishes of the genus *Eleotris* found in Japan," Japanese Journal of Ichthyology, Vol. 14, pp 135–166, 1967. https://doi.org/10.11369/jji1950.14.135

"Additional research on the scapula of gobiid fishes," Japanese Journal of Ichthyology, Vol. 14, pp 167–182, 1967. https://doi.org/10.11369/jji1950.14.167
"A systematic examination of the gobiid fishes based on the mesoptergobid, postcleithra, branchiostegals, pelvic fins, scapula, and suborbital," Japanese Journal of Ichthyology, Vol. 16, pp 93–114, 1969. https://doi.org/10.11369/jji1950.16.93
"On the supratemporals of gobiid fishes," Japanese Journal of Ichthyology, Vol. 18, pp 57–64, 1971. https://doi.org/10.11369/jji1950.18.57
"On a specimen of "Matsugehaze," *Oxyurichthys ophthalmonema*, collected in Kanagawa Prefecture, Japan," Japanese Journal of Ichthyology, Vol. 19, pp 103–110, 1972. https://doi.org/10.11369/jji1950.19.103

The papers co-authored with Katsusuke Meguro are listed chronologically below:

"On gobiid *Ophiocara porocephala* and *Ophieleotris aporos*," Japanese Journal of Ichthyology, Vol. 21, pp 72–84, 1974. https://doi.org/10.11369/jji1950.21.72
"First record of *Glossogobius celebius* from Japan," Japanese Journal of Ichthyology, Vol. 21, pp 227–230, 1975. https://doi.org/10.11369/jji1950.21.227
"On gobiid fish, *Cryptocentroides insignis* from Okinawa Prefecture, Japan," Japanese Journal of Ichthyology, Vol. 21, pp 231–232, 1975. https://doi.org/10.11369/jji1950.21.231
"On a goby *Redigobius bikolanus*," Japanese Journal of Ichthyology, Vol. 22, pp 49–52, 1975. https://doi.org/10.11369/jji1950.22.49
○"*Pandaka trimaculata*, a new species of dwarf goby from Okinawa Prefecture and the Philippines," Japanese Journal of Ichthyology, Vol. 22, pp 63–67, 1975. https://doi.org/10.11369/jji1950.22.63
"On a goby *Callogobius okinawae*," Japanese Journal of Ichthyology, Vol. 22, pp 112–116, 1975. https://doi.org/10.11369/jji1950.22.112
○"Description of a new gobiid fish, *Glossogobius aureus*, with notes on related species of the genus," Japanese Journal of Ichthyology, Vol. 22, pp 127–142, 1975. https://doi.org/10.11369/jji1950.22.127
○"*Glassogobius sparsipapillus*, a new species of goby from Vietnam," Japanese Journal of Ichthyology, Vol. 23, pp 9–11, 1976. https://doi.org/10.11369/jji1950.23.9
"Five species of the genus *Callogobius* found in Japan and their relationships," Japanese Journal of Ichthyology, Vol. 24, pp 113–127, 1977. https://doi.org/10.11369/jji1950.24.113
"First record of the goby *Mangarius waterousi* from Japan," Japanese Journal of Ichthyology, Vol. 24, pp 223–226, 1977. https://doi.org/10.11369/jji1950.24.223
"First record of the goby *Myersina marcrostoma* from Japan," Japanese Journal of Ichthyology, Vol. 24, pp 295–299, 1978. https://doi.org/10.11369/jji1950.24.295
"On the differences between the genera *Sicydium* and *Sicyopterus* (Gobiidae)," Japanese Journal of Ichthyology, Vol. 26, pp 192–202, 1979. https://doi.org/10.11369/jji1950.26.192
"On the six species of the genus *Bathygobius* found in Japan," Japanese Journal of Ichthyology, Vol. 27, pp 215–236, 1980. https://doi.org/10.11369/jji1950.27.215

"A gobiid fish belonging to the genus *Hetereleotris* collected in Japan," Japanese Journal of Ichthyology, Vol. 28, pp 329–339, 1981. https://doi.org/10.11369/jji1950.28.329

○"*Myershina nigrivirgata*, a new species of goby from Okinawa Prefecture in Japan," Japanese Journal of Ichthyology, Vol. 29, pp 343–348, 1983. https://doi.org/10.11369/jji1950.29.343

○"Two new species of goby of the genus *Astrabe* from Japan," Japanese Journal of Ichthyology, Vol. 34, pp 409–420, 1988. https://doi.org/10.11369/jji1950.34.409

It should be noted that, at the "Second International Conference on Indo-Pacific Fish," held in 1986, the then-Crown Prince made an oral presentation in English of his paper entitled

◎ "Some morphological characters considered to be important in gobiid phylogeny," which was printed in the "Proceedings."

By the way, the then-Crown Prince served as the Honorary President of the Conference.

The papers marked with a prefix ○ are ones in English with a Japanese abstract, those marked with a prefix ◎ are in English only, and the others are in Japanese with an abstract in English. The legends of all of the figures and tables in his papers are described in English.

His co-authored papers written under the name Akihito after his accession include:

"Reexamination of the status of the striped goby," Japanese Journal of Ichthyology, Vol. 36, pp 100–112, 1989 (co-authored with Katsuichi Sakamoto) https://doi.org/10.11369/jji1950.36.100.

○"Review of the gobiid genus *Cristatogobius* found in Japan with a description of a new species," Ichthyological Research, Vol. 47, pp 249–261, 2000 (co-authored with Katsusuke Meguro) https://doi.org/10.1007/BF02674248, and

○"A new species of gobiid fish, *Cristatogobius rubripectoralis*, from Australia," Ichthyological Research, Vol. 50, pp 117–122, 2003 (co-authored with Katsusuke Meguro) https://doi.org/10.1007/s10228-002-0147-1.

These papers discuss the taxonomy and phylogeny of gobiid fish, mainly in terms of their morphological features. In recent years, he has further expanded his interest to the elucidation of the molecular phylogeny of gobies based on DNA sequences, and has published the 3 papers listed below jointly with a group led by Takashi Gojobori, working at the National Institute of Genetics, Research Organization of Information and Systems. As described later, Gojobori was Prince Akishino's supervisor when he submitted his thesis on the phylogeny of domestic fowls. Yet, the first of the papers listed below was co-authored with Susumu Ohno, who advocated the theory of "gene duplication in evolution," and the second was done with Masami Hasegawa, working at the Institute of Statistical Mathematics (ISM) (also a subgroup of the Research Organization of Information and Systems), who first suggested that humans (genus *Homo*) and chimpanzees (*Pan troglodytes*) are in close genetic relation to each other.

◎ "Evolutionary aspects of gobioid fishes based upon a phylogenetic analysis of mitochondrial cytochrome b genes," Gene, Vol. 259, pp 5–15, 2000 (co-authored with the Prince Akishino, Akihisa Iwata, Yuji Ikeda, Katsuichi Sakamoto, Takashi Gojobori, et al.) https://doi.org/10.1016/S0378-1119(00)00488-1.

◎ "Evolution of Pacific Ocean and the Sea of Japan populations of the gobiid species *Pterogobius elapoides* and *Pterogobius zonoleucus* based on molecular and morphological analysis," Gene, Vol. 427, pp 7–18, 2008 (co-authored with the Prince Akishino, Yuji Ikeda, Masahiro Aizawa, Tetsuji Nakabo, Takashi Gojobori, et al.) https://doi.org/10.1016/j.gene.2008.09.026.

◎ "Speciation of two gobioid species, *Pterogobius elapoides* and *Pterogobius zonoleucus*, revealed by multi-locus nuclear and mitochondrial DNA analyses," Gene, 576, pp 593–602, 2016 (co-authored with Prince Akishino, Yuji Ikeda, Masahiro Aizawa, Masami Hasegawa, Tetsuji Nakabo, Takashi Gojobori, et al.) https://doi.org/10.1016/j.gene.2015.10.014.

In addition to these researches, he made a study of the life mode of the raccoon dogs that lived on the grounds of the Imperial Palace with Takako Sako, et al. The results of the study were compiled into a paper and published:

"Seasonal food habits of the raccoon dog, *Nyctereustes procyonoides*, in the Imperial Palace, Tokyo," Bulletin of the National Science Museum Series A (Zoology), Vol. 34 (II), pp 63–75, 2008.

One of the features in the lifestyle of raccoon dogs has long been known as "Tanuki no tameguso." This expression sprang from their habit of multiple individuals all excreting at the same site. For this reason, a research on a pile of excretion left at the site allowed us to determine their feeding habits and the number of individuals within a given area. It is said that the Emperor examined the excretions himself.

Looking back at his achievements in this way, since publishing his first original paper in 1963 in his Crown Prince days, he has presented papers on his research on gobies almost every year, and, in 1975 alone, he wrote as many as 7 papers with Katsusuke Meguro. In 1975, he wrote the following poems:

> I named a goby collected in the sea area around Iriomotejima situated in the southern part of Japan, "*Kuro'obi-haze*" (*Myershina nigrivirgata*), and recorded the name as a new species.
>
> I think of the future of the southern island, where new species of goby have been discovered one after another.

As can be seen from this waka poem, he seemed to feel fulfilled by a life studying gobies. However, he had also been very busy with his various official duties for a certain period of time, from Emperor Showa's late years to several years after his accession to the throne. Accordingly, he could publish only 6 papers after his accession, because of the constant need to give top priority to his official duties. At the press conference held for the celebration of his birthday in 1998, he provided the following answer to the question, "If you had some time, what would you like to do?":

Fig. 2.12 Fish of Japan with Pictorial Keys to the Species, to which Emperor Akihito contributed the section on gobioid fishes (Courtesy of the Imperial Household Agency)

> I would like to continue writing papers on gobies. In my Crown Prince days, I could write papers almost every year, but after my accession to the throne, I could no longer do so because of being too busy with my official duties, save for a single paper written in 1989, the year of accession. I feel that time flew by so fast after my accession, while, in truth, I have been away from research for a long time. This summer, I could resume writing my paper and make a little progress with the work. In the course of writing, I could feel happy, even while experiencing difficulty in determining the truth.

This statement was confirmed by the fact that, two years later, he published a further original paper, as well as co-authoring the section describing gobies from "Fishes of Japan with Pictorial Keys to the Species," edited by Tetsuji Nakabo and published by Tokai University Press in 1993. A second pressing of the first edition, revised and enlarged, was published in 2000, two years later, the English edition in 2002 and the third edition in 2013. The third edition contained a great many sketches, and amounted to over four hundred pages in total (Fig. 2.12). The books and research papers that he wrote in the fields of biology and the natural sciences are listed below:

"Evolution of gobiid fish," Kagaku Salon, Vol. 8, No. 3, 1984, Tokai University Press

"The Fishes of the Japanese Archipelago, Suborder Gobioidei" (co-authored with Masayoshi Hayashi, et al.), first edition: 1984, Tokai University Press; second edition: 1988, Tokai University Press

"On Japanese Freshwater Fishes—Distribution, Mutation and Speciation," Dusky tripletooth goby (*Tridentiger obscurus*) (co-authored with Nobuhiko Mizuno, et al.), 1987, Tokai University Press

The Emperor Akihito's Manuscript ("Evolution of gobiid fishes"), Bungeishunju, Vol. 77, No. 10, 1999

"Tennoheika Kagakuwokataru" ("The Emperor Talks About Science"), Asahi Shimbun Publications, 2009 [4].

Highlights and Evaluation of Emperor Akihito's Research and Writings

In recent years, in the fields of biology and the life sciences, in which the papers accepted by and then published in academic journals such as "Nature," "Science" and "Cell" are highly esteemed, researchers all over the world usually make efforts to swim with such a tide. However, in some fields, experts more frequently read specialized journals relevant to their subjects than the non-specialized journals; giving an example, in the field of mathematics, the papers published in bulletins, which are issued by the universities, may be more highly esteemed worldwide than those in the journals. The research papers in the fields of taxonomy and phylogeny may contain a lot of sketches, or may be voluminous, or may have to be distributed widely. For this reason, the researchers tend to contribute to, in the case of fishes, the "Japanese Journal of Ichthyology" or the bulletins issued by universities and institutes. Thus, these papers are appreciated at the same level as those in other fields, both in terms of quality and quantity.

The Emperor's basic research on the new species of goby that he himself described and his own classification based on the scapula of gobiid fishes have been described earlier. One of his most noteworthy achievements is "On four species of the gobiid fish of the *Eleotris* found in Japan." Gobies have sensory organs on their heads lined up in both the longitudinal and lateral directions, called head pit organs. These sensory organs transfer low-frequency vibrations through water. He revealed that the arrangement pattern of the head pit organs is a singular trait (the morphological, physiological and functional characteristics of organisms) of gobies, which enables researchers to discriminate among 4 species of the genus *Eleotris* in Japan and two species of same in Madagascar and Hawaii, through a close examination of various kinds of morphological feature for the first time in history. Initially, some expressed doubt regarding this method of discrimination, but, later, it was gradually recognized and has become commonly used, pushing toward the establishment of goby taxonomy. He reminisces that:

> Looking back, I remember feeling great pleasure when I successfully clarified the features of the pit organ arrangements of the individual species of gobies while continuing to watch them under a microscope. At present, the arrangement of the pit organs serves as an important base for classifying gobioid fishes. I am pleased that I could make some contribution to this field in this way [4].

As his second achievement, he made a presentation in English of "Some morphological characters considered to be important in gobiid phylogeny" at "The 2nd Indo-Pacific Fish Conference," held in 1986. Two years before that presentation, he had written an article entitled "Evolution of gobiid fish." The group of gobies, for which no sufficient evidence of a fossil record has been provided and yet are rich in variation, has a relatively short history of evolution. Accordingly, they are the most suitable for research on phylogenic evolution. He made the presentation as a final compilation of the results of his morphological research accumulated so far. In that paper, he divided the Gobiidae into the two subfamilies, Eleotrinae with perfectly separated pelvic fins and Gobinae with not separated pelvic fins. He concluded that, overall, the process of losing the bones and sensory organs could be seen, and that *Bostrychus sinensis* from Japan and *Oxyeleotris marmorata* from Southeast Asia of the subfamily Eleotrinae could be considered to represent the form closest to the ancestral form. Partially because he made the presentation in English, his research has become further widely known throughout the world.

His third achievement is the successful phylogenetic classification of gobies using mitochondrial DNA analysis, which was published in "Gene" in 2000. He was compelled to neglect his research activity due to his hectic official duties during the period before and after his accession, while great strides were being made in the field of research on gobies. This work was started as a joint research with Prince Akishino when the Prince introduced Takashi Gojobori, a specialist in this area and one of his academic advisors, to the Emperor. It is a rare case that a father and his child, even members of a family line of scientists, can find such opportunities to discuss research on a specific theme in the same academic field. Even though this kind of research has gradually become more popular in the world, there is no doubt that they took the initiative in doing such work and produced the historical achievement of the phylogenetic classification of gobies involving discussions from the aspect of morphology. Around the same time, I also had an experience with the phylogenetic classification of butterflies, but morphologists were unwilling to consider the results of the molecular analysis. It is a rather rare case that the Emperor took such a position that both molecular phylogenetic and morphological results could flexibly be combined for consideration.

In his paper published in "Gene" in 2008, he undertook a work to discriminate between the Pacific Ocean and the Sea of Japan populations of the gobiid species, *Pterogobius elapoides*, based on a difference in the number of stripes on the side of the body. However, he was not satisfied with the sampling *P. zonoleucus*, closely related to *P. elapoides*, as a reference species collected at only two points and made a suggestion that it might be better to collect *P. zonoleucus* at a far greater number of points, taking this opportunity to do so (Fig. 2.13). According to his suggestion, an increase in the sampling points revealed a difference between the Pacific Ocean and the Sea of Japan populations of *P. zonoleucus*, which are apparently difficult to discriminate from each other. Furthermore, this work elucidated the process by which the *P. zonoleucus* and *P. elapoides* populations of the Sea of Japan and the *P. zonoleucus* population of the Pacific Ocean initially branched off from their common ancestor and, in turn, the *P. elapoides* populations of the Pacific Ocean and the Sea of Japan have branched off from the latter. It also manifested the possibility that the *P. zonoleucus* population of the Sea of Japan might be a new species. In his later paper published in "Gene" in 2016, he discusses the process of the above-mentioned speciation, using not only mitochondrial DNA, but also nuclear DNA.

Needless to say, all of these papers were co-authored with the authoritative people, including Gojobori, as well as Susumu Ohno and Masami Hasegawa. However, this kind of research usually tends to be conducted jointly by multiple researchers, and the Emperor is the only person who deserves the status of primary

Fig. 2.13 *Pterogobius
elapoides* and *P. zonoleucus.*
From uppermost, *P. elapoides*
population of the Pacific
Ocean, *P. elapoides*
population of the Sea of Japan
and *P. zonoleucus* population
of the Pacific Ocean (From
"Fishes of Japan with
Pictorial Keys to the Species,
3rd edition" courtesy of
Akihito el al.)

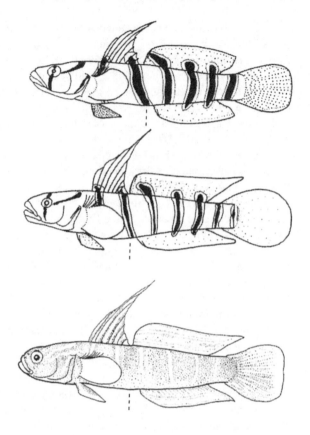

author in terms of the papers' high level of quality. Furthermore, the 3rd paper was
published when he was 82 years old. It is extremely difficult for ordinary persons to
obtain such good results at that age.

Peter J. Miller, a professor at the University of Bristol, who had academic
interaction with the Emperor, states in "The Emperor and Taxonomy of Gobies"
that, in conjunction with said interaction, he enjoyed the benefits of the Emperor's
achievements:

"The Emperor began his research on the taxonomy of gobies while he was the
Crown Prince and published his first paper, entitled "On the scapula of gobiid fish,"
in 1963. Then, he not only published 26 papers up until the publication of the paper
"Reexamination of the status of the striped goby" in 1989, but also contributed
articles to pictorial books of fish, scientific journals, and so on. No one is more
accomplished as a scientist than him among the members of European royal
families. The Emperor's papers, which are characterized by strict and
highly-detailed descriptions of species … have made a substantial contribution to
the researches of other taxonomists of gobies. … The results of the Emperor's
researches were compiled into the descriptions of 200 species of gobies from Japan
in the book "The Fishes of the Japanese Archipelago." They have provided

extremely valuable information to researchers; in particular, information on the arrangement pattern of the neural cells of the pit organ is not readily available from any other sources of information (Re-translated from the Japanese text [13]).

Tetsuji Nakabo, who asked the Emperor to contribute his manuscript to "Fishes of Japan with Pictorial Keys to the Species," and who also, at present, continues to provide advice on the Emperor's research, recognizes the Emperor as a superior scientist who steadily achieves his results through a rigorous approach, stating:

> He neither speaks nor writes about anything in his papers that is beyond reason or scientifically unproven. Moreover, he does not readily trust anything that he has not seen himself, and always employs a stringent system of verification until he is completely satisfied. In any case, he never cuts corners. I really respect him for his sincere attitude.

Furthermore he has written in an article:

"The Emperor conducts his research on morphological classification with his chamberlains and specialists, but, first of all, he observes specimens and writes papers on the results of said observation by himself. Although he has co-authored papers on phylogenetic research with other researchers, including me, he is always the primary author on these papers, because he proposes the issue to be treated as a subject of research, he determines the direction of the research and he paves the way leading to its conclusion. The number of papers written by him as primary author far exceeds those of other ordinary researchers. Considering that, unlike us professional ichthyologists, he takes time out of his busy schedule of official duties to conduct his research, I have to take my hat off to him for his sincere enthusiasm" [53]. Nakabo, in a letter to me, wrote, "I have had the honor of meeting with the Emperor to discuss his research once a month for about 5 years, and I would be so irreverent as to say that his sensibility in regard to research can be summed up in one word: 'fresh.'" It is said that the eponyms derived from Emperor Akihito include 3 for gobiid species found in the sea around Japan and 1 species of 1 genus (*Akihito vanuatsu*) from the sea of a foreign country.

His taxonomical research on gobies is highly esteemed abroad, as evidenced by his appointment, in his Crown Prince days, as a foreign member of the Linnean Society, an honor that has been bestowed upon only 50 scientists in the world, based on their academic achievements. By the way, Emperor Showa was also an honorary member of the Society. At that time, only Hiroshi Hara, a botanist, had been appointed as a member from Japan. At present, no names of Japanese scientists can be found in the list of foreign members. Emperor Akihito was recommended for an honorary membership in the Society in 1986.

Moreover, he was appointed as an honorary research associate of the Australian Museum, an honorary member of the Zoological Society of London and a permanent honorary member of the Research Center of the Natural Sciences in Argentina. Additionally, as described earlier, he received the King Charles II Medal.

His Profound Knowledge of the History of Science in Japan

"Science," an American scientific journal, was planning to release a special issue all about Japan in 1992. At that time, Emperor Akihito contributed his over 3,000 word article in English, entitled "Early Cultivators of Science in Japan" (Vol. 258, pp 578–580, 1992), in response to a request from the editor through the intermediary of Takashi Sugimura, the Honorable President of the National Cancer Center (the former President of the Japan Academy). Such a request from 'Science' provides the most conclusive evidence that the Emperor is more highly esteemed as a superior scientist abroad than at home. The content of the article was apparently left to him; thus, he thought about the fact that the historical background and process that led to Japan's contributions to the international scientific field in recent years were not widely known around the world. This is why he selected this theme.

Starting with the introduction of guns to Japan, the article describes the way in which the study of advanced sciences developed in European countries through the Dutch Trading House in Dejima, Nagasaki, during the period of national isolation (1639–1854); the efforts towards and significance of the translation of "Kaitai Shinsho" ("A New Book of Anatomy") from Dutch to Japanese by Genpaku Sugita, et al.; the achievements of Gen'etsu Kagawa, an obstetrician who contributed to an improvement of medical science in Japan through his unique research; the exchanges of Japanese scientists with the physicians Carl Peter Thunberg and Philip Franz von Siebold, who visited Japan and worked at the House; the performance of Yozo Yamao, who made a strong recommendation for the establishment of an engineering school, stating his philosophy that "Even if there is no industry at present in Japan, if we were to train a man, he would promote the industry," and the contribution of Yukichi Fukuzawa, an enlightenment thinker and educator in the Meiji Period, ultimately concluding:

> Another 100 years have passed since then, and it is a matter for rejoicing that, thanks to the efforts of many scientists, science in Japan has made steady progress and has become able to contribute to the world's scientific community. At the same time, I cannot help recalling, with a sense of gratitude and respect, those people who exerted their untiring efforts for the development of science in Japan at its infant stage, under the severe conditions of national isolation, without a teacher and relying solely on books that were brought from Europe (translated from the Japanese article in "Bungeishunjyu", Special Issue of October, 1999 [13]).

Shuntaro Ito of the University of Tokyo, who is an authority in both the history of science and the philosophy of science, states, in an article contributed to the same issue [13]:

> The Emperor's words impressively convey to foreign scientists and intellectuals the road that was filled with hope, as well as being marked with difficulties, along which Japanese intellectual pioneers walked to absorb and digest the sciences developed in Europe by means of the Dutch language during that period (*starting from the introduction of guns up to the Meiji Restoration). Moreover, he describes Gen'etsu Kagawa, the founder of modern obstetrics in Japan, and Yozo Yamao, the Minister of Engineering in the early Meiji Period, both of whom usually get no credit, making this a unique statement by the Emperor.

Ito emphasized that the Emperor's article in "Science" was an original work, conceived and executed through the consultation of academic literature and other materials. The one negligible exception was that Ito had had a chat with the Emperor about drafting the article. The same is true of the "okotoba" ("messages") that he addressed at various opportunities. The Emperor listens to the opinions of specialists, consults specialized literature and then composes the messages or writes the papers in his own words.

Just like Emperor Showa, Emperor Akihito has been interested in history and biology since he was a child, as described earlier. Although writing modestly in the "Science" article that, "I have made no research on the history of science," he actually has a deep knowledge of the history of biology, including the classification of gobies, as well as the general history of science in Japan. In his keynote speech, "Linné and Taxonomy in Japan—On the 300th Anniversary of his Birth," in 2007, he not only detailed the exchanges between Carl Peter Thunberg, Hoshu Katsuragawa and Jun'an Nakagawa, as well as those between Philipp Franz von Siebold and Keisuke Ito, starting from Engelbert Kaempfer, but also those among the hired foreign teachers in the Meiji Period. In addition, he dealt with the discoveries of the spermatozoids of ginkgo by Sakugoro Hirase and the spermatozoids of cycad by Sei'ichiro Ikeno. In the latter half of his speech, while talking about the taxonomy of gobies, he did not forget to touch on the difficulties experienced by most Japanese researchers, in short, the fact that they were forced to use the type specimens stored in foreign laboratories and museums. It is worthy of praise that he emphasized the last point. This speech, entitled "Linnaeus and taxonomy in Japan," was published in "Nature" in 2007 (Vol. 448, pp 139–140, 2007).

Besides, he often delivered talks about the history of science in individual scientific fields in Japan in his "okotoba." For instance, at the opening ceremony of the general meeting of the International Astronomical Union held in 1997, he spoke about Harumi Shibukawa, who created a calendar unique to Japan in the 17th century, and Hisashi Kimura, who was famous for adding the Z term, as a correction term, to the formula concerning the axis of the rotation of the earth and was the first to be presented with the Order of Culture, while referring to the history of astronomy, ranging from astrology to artificial satellites. As is well known, the above-mentioned bi-generational interests in history and biology have been passed down to the present Emperor's children, the former to Crown Prince Naruhito, and the latter to Prince Akishino and Sayako Kuroda.

Flora and Fauna of the Imperial Palace

Finally, we cover the Emperor's contribution to the survey of organisms living in the Imperial Palace. In his message delivered at the award ceremony for the International Prize for Biology in 1988, he made own thoughts clear, suggesting the importance of nature conservation:

During the long history of life on this planet, many species were born and many have died out. Also, what is so saddening is the realization that in the relatively short span of time that humans have inhabited this Earth, an inordinate number of species have been driven to extinction by human activities and the impact of these activities on the environment. We shall never again witness the flight of the passenger pigeon which once darkened the skies with its sheer numbers, nor can we ever again see in living form the flightless great auk which occupied an ecological niche in the northern seas much like that of the penguins in the southern seas. The coutless specimens of extinct birds that I saw at the British Museum, Natural History forty-five years ago have left a deep and indelible impression on me.

As described repeatedly, the Fukiage Imperial Garden in the Imperial Palace has been restored to its current appearance, with a forest and wild herbs that most closely resemble what would naturally occur there, according to the intention of the Emperor Showa, through cessation of any trimming of the garden, including mowing and transplanting various kinds of plants from the external environment, especially from Musashino. Along with the restoration, many species of animal have come to live there. In the midst of the multi-story reinforced-concrete buildings that stand together in large numbers, the Imperial Palace, including the East Imperial Garden, forms a unique space for a wide variety of species of animals and plants to make their habitat. Needless to say, the Emperor has deep knowledge of these animals and plants and walks around the Garden with Empress Michiko, just like Emperor Showa and Empress Kojun. Looking at the white birch trees and yu'usuge (yellow daily lily, *Hemerocallis citrina var.vespertina*) transplanted from Karuizawa, they seem to recall their younger days. By the way, the Empress Michiko's official symbol is a white birch tree. Moreover, Hamagiku (Nippon daisy, *Nipponanthemum nipponicum*) opens up in autumn in the Garden. Its seeds had been presented by the proprietor of a hotel situated in Ozuchi-cho, Iwate, who passed away in the Great East Japan Earthquake in 2011. The hotel has been successfully restored by the successors, who got the courage to do so from the photo of Hamagiku in the Imperial Palace. Their Majesties visited the hotel five years later.

As described earlier, at the beginning of the Heisei Period, the Emperor suggested that it might be of great significance to conduct a scientific survey and research on the flora and fauna on the grounds of the Imperial Palace and then record the results, and also recommended that any changes over time be examined several years later. In response to his suggestion, the researchers from the universities and institutes throughout the country, centering on the Natural Museum of Nature and Science, jointly conducted a survey around the Fukiage Imperial Garden and the Docan-bori moat during the period from 1996 to 2001, recording 1,366 species of plant and 3,638 species of animal (quoted from the book "Organisms Living in the Imperial Palace and Fukiage Garden" [57]), etc. It is said that he also observed birds through a telescope with the Empress and the Princess Sayako, and that, at night, he surveyed the scene at the site where the researchers were conducting a survey on bats and moths. The survey on the flora and fauna on the grounds of the Imperial Palace was again conducted by the National Museum of Nature and Science during the period from 2009 to 2013, because the record of the

Fig. 2.14 Emperor Akihito and Empress Michiko walking along the nature observation route in the Fukiage Gardens, scheduled to be open to the public in 2007 at the time (Courtesy of the Imperial Household Agency)

changes that took place over time might have been of significance, and its results were published in 2014. As many as roughly 200 researchers participated in the survey. In the latter survey, 250 species of plant and 649 species of animal were added to the previous record, and, among them, about 100 organisms, including a wild flower, "Fukiage nirinso (*Anemone imperialis*)," and a sawfly, "Nihon-koshiaka-habachi, (*Siobla japonica*)," thought to be new species, were included. According to Takako Sako of the Biological Laboratory, more than 10 raccoon dogs, as well as an unknown number of masked palm civets (*Paguma larvata*), most probably live in the Garden. This finding is an admirable achievement, comparable to the Emperor Showa's "Flora of the Imperial Palace, Tokyo" and "Flora of the Nasu Imperial Villa," or the sample collection and examination in Sagami Bay (Fig. 2.14).

According to Empror Akihito's intention, a number of plants have been planted in the East Imperial Garden, which is open to the public, as already described, and an identification tag describing the unique Japanese, family and Latin scientific names was placed by each of these plants. Several years ago, these tags were replaced with new ones for easier identification, with the first replacements being done by Their Majesties themselves. The Emperor bore the expense of the replacement. To try to conserve old breeds of fruit trees, he created the "Kaju Kohinshu Garden" (an orchard of old breeds of fruit trees) in the East Imperial Garden and planted several tens of pear, peach, kaki (persimmon) and apple trees there, all of which are obsolete breeds (quoted from the book written by Makoto Watanabe [68]). As described a little later on, in addition to evoking historical interest and a sense of values, the fruit trees may have opportunities to be useful in the future. Furthermore, in the Omichi Garden, which is closed to the public, bonsais (dwarfed trees in a pot), flowers and ornament plants are being grown to decorate the Palace.

Except for the members of labor service groups, who provide cleaning services, no one is usually allowed to enter the Imperial Palace and the Akasaka Imperial Residences. In 2007, "Midori no Gekkan" ("Greenery Month"), which takes place from April 15 to May 14, was established. With this event as a start, the Emperor hoped that a lot of people might enjoy the natural environment in the Fukiage Imperial Garden and, at his instruction, a nature observation event open to the public on Greenery Day (May 4), Children's Day (May 5) and November 15 (at present, only April and May) has come to be held. The researchers at the National Museum of Nature and Science, who were involved in the examination of the flora and fauna in the Garden, are on hand to explain the organisms. Initially, even though the quota was 200, as many as 40,000 persons applied for admission. For the first 5 days of April 2014, the Imperial Palace was opened to the public to celebrate the Emperor's 80th birthday, and more than 380,000 people visited there to see the cherry blossoms in full bloom. For the first 5 days of November, it was also opened to the public for them to enjoy the autumn foliage, and 350,000 people visited. These events have provided the people with their best opportunity to enjoy the Palace, and it is all because of the Emperor's enthusiasm to have the people enjoy all of this beauty along with him.

It has been already described that rice-planting and rice-harvesting in the paddy field, a part of the Biological Laboratory, were events initiated by Emperor Showa. Emperor Akihito, of course, has inherited these events, and is known to work in the fields himself, wearing work clothes and high boots. Moreover, he has worked on sowing rice seeds in the bed for rice seedlings. In addition to the rice seeds, which are sent by the Institute of Crop Science, National Agriculture Food Research Organization, every year, he also sows the rice seeds harvested in the previous year. Inevitably, the amount of rice seeds to be sown significantly increases, making his workload, at present, as much as 20 times that of the initial year. To avoid overload, the sowing of the rice seeds harvested since 2009 was discontinued. The reason why he increased the amount in the first place is that he had a biological interest in the possibility that it might produce information on the variety. The rice-harvesting work is carried out by all of the Imperial family members, including the Prince Hisahito, one of his grandchildren, in autumn of every year (Fig. 2.15).

At the Momiji-yama Hill Sericultural Station, where the Empress works on silkworm cultivation, an old breed of silkworm called "Koishimaru" has been grown in small quantities. These silkworms can only produce cocoons of a small size and finer silk threads; accordingly, other breeds of silkworm have increasingly come to be grown and have taken the place of the old ones, which had barely been reared. In regard to this situation, Empress Michiko said:

Fig. 2.15 Emperor Akihito sowing rice seeds in the paddy field near the Biological Laboratory in 2014

I have heard that they are Japanese purebred silkworms and that the shape of the cocoons is pretty, and that their threads are fine and very beautiful. I hope to make the old breed survive so that we can continue to grow Koishimaru for the near future.

Then, while the work, started in 1994, was taking place on restoring and making replicas of the silk fabrics, among the Imperial properties preserved in the Shosoin, it was revealed that, out of various kinds of thread, the ones produced by Koishimaru were closest to those produced in the Nara Period, during which the Shosoin was built. The Empress reared a large number of Koishimaru, equivalent to several times the previous batch, to supply enough silk to restore the Imperial properties and weave their replicas. This work was successfully completed after ten years. She grew a still greater number of Koishimaru to use for the silk fabrics required in the "Sengu" event (transfer of a deity to a new shrine built every 20 years) of the Ise Shrine (Makoto Watanabe [68]) (Fig. 2.16).

Breeds of both domestic animals and edible plants have been improved so that the traits that are useful to human life may be attained, while many genes have been lost. However, those lost genes may contain the codes that enable resilience to rapid

Fig. 2.16 Empress Michiko feeding silkworms with mulberry leaves at the Momiji-yama Hill Sericultural Institute in 2007 (Courtesy of the Imperial Household Agency)

climate change and resistance to plagues. In short, the loss of even seemingly useless genes could lead to the extinction of all species and all of their varieties. By the way, the case of the Koishimaru demonstrates how important the conservation of gene resources is.

With the development of the sericultural industry from the beginning of the Edo Period, many breeds of silkworm were created and the number of distributors trading in silkworm eggs increased in Japan. However, the Japanese sericultural industry began a steady decline in 1969, when China superseded Japan's first-in-the-world position in the silk thread and silk fabric industries. In line with the declination, the number of silkworms also decreased, and they are now almost extinct. Although silkworms had provided geneticists and embryologists with precious materials since the rediscovery of Mendel's law in animals by Kametaro Toyama (1906), it became difficult to sustain the various breeds of silkworm necessary for researches. It is said that, at present, Kyushu University is playing a central role in making efforts to collect and conserve them.

Prince Hitachi's Personal History and His Research on Cancer

Prince Hitachi Masahito was born in 1935, and was called "Prince Yoshi" in his childhood. Just like Emperor Akihito, he left his home at the age of 4 and was raised by more than one fuikukan (a tutor in charge of the Imperial Prince) at a palace different from the one where the Crown Prince was raised. The chief fuikukan was Hidesaburo Koori (a vice admiral). Torahiko Nagazumi, one of the Emperor Showa's fellow students, had long worked at the Prince Yoshi's Palace, and Motobumi Higashisono, a fuikukan for the Crown Prince, also served him. He suffered a minor bout of polio when he was a very young child, but underwent rehabilitation for about 2 years, fully recovering from the disease. At the end of World War II, he evacuated to a rural region, just like Emperor Akihito. After the war, he became widely familiar to the public under the nickname "Kasei (Mars)-chan." He is known for being an academic-type prince, just as his father was, and his facial features even resemble Emperor Showa's.

He has loved living organisms since his childhood, and often accompanied his father in collecting marine organisms. He also always helped his father in collecting plant specimens at Nasu. When entering the university, he was far freer to select his desired major than the Crown Prince had been. However, unfortunately, no biology department had yet been set up at Gakushuin University at that time, and so he entered the Department of Chemistry, Faculty of Science, as his second choice. To satisfy his strong attraction to biology, Takashi Fujii, a professor, and others working at the Department of Zoology, University of Tokyo, visited his palace to deliver their lectures individually. Because of that, he was able to realize his original intention and, in 1958, after graduating from Gakushuin University, he entered Professor Fujii's Laboratory as a research student. As a precedent, it is recalled that the Prince Mikasa, the youngest brother of the Emperor Showa, entered the Faculty of Literature, University of Tokyo, as a research student to walk a path of studying oriental history after the war.

Takashi Fujii, who is an embryologist and was 4 years behind Ichiro Tomiyama, delivered the lectures of the second Laboratory, "Histology" and "Embryology." Later, he was appointed as the dean of the Faculty of Science and served as an acting president for a period during the Yasuda Auditorium Incident in 1969, until Ichiro Kato assumed the presidency. Initially, Takeo Mizuno and Sei'ichiro Kinoshita, who were research associates at that time (both were later appointed as professors at the University of Tokyo), guided him. Fujii's laboratory was on the 2nd floor of the antique-looking No. 2 building of the Faculty of Science near "Akamon" (a red gate situated at the southwest corner of the university's campus). He would often visit the Misaki Marine Biological Station, together with Chamberlain Higashisono and others, to conduct practical works and experiments when I worked there.

He acquired a wide range of knowledge on developmental biology and, specifically, did research on the cell division of the crop sac epithelium of pigeons.

Pigeons, both male and female, have a physiological mechanism by which the endothelium of their crop sac on an esophageal wall proliferates under the influence of a hormone, prolactin, excreted from the pituitary gland, producing a kind of milk that they give to their chicks. He extensively studied the mechanism of cell division that occurs when the epithelium proliferates. In 1959, his first paper,

"The effect of nicotinamide on prolactin-induced mitosis of crop-sac epithelium in the pigeon," Journal of the Faculty of Science, University of Tokyo, Vol. 8, pp 418–417,

was published. After that, he published 3 more papers in this journal, reporting the results of a series of researches on this theme, over the course of several years. He was always genial-looking, and communicated comfortably with the young researchers of both zoology and botany in the same building.

As he was also interested in insect-collecting, he visited Amami-Oshima to collect butterflies, together with Minoru Osanai (later appointed as a professor at Kanazawa University), an upper-class student studying at the laboratory who revealed the ecology of the white letter hairstreak (*Fixsenia w-album*). Then, just after returning home from his overseas trip, he took time to put in a call to Osanai at his home, so that he could tell him the names of the butterflies that he had collected. He also put the larvae of the great purple emperor, Japan's national butterfly (*Sasakia charonda*), on a Chinese hackberry tree (*Celtis sinensis*) on the grounds of the Imperial Palace to be raised. Furthermore, being interested in birds as well as insects, he often went out bird-watching, together with Yasuo Uchida, who was his junior by a few years and who became an expert in ornithology (later appointed as a professor at Surugadai University). Around that time, he succeeded Nobusuke Takatsukasa and Yoshimasa Yamashina in serving as the President of the Japanese Society for the Preservation of Birds.

Meanwhile, in 1964, a wedding was arranged with Hanako, a daughter of the ex-lord of the Tsugaru Domain (the current Aomori), and newly founded the Prince Hitachi Family, the first Prince Family to be founded after the war. However, in 1968, the Yasuda Auditorium Incident occurred, and the No. 2 building of the Faculty of Science, which housed the laboratory that he and other researchers used, was occupied by campus activists. For this reason, early in the following year, he transferred, as a research fellow, to the Cancer Institute, the Japanese Foundation for Cancer Research (JFCR), situated at Otsuka, Tokyo, of which Tomizo Yoshida, who was familiar with Fujii, was president at that time. If the university dispute had not occurred, he would have continued his research at the University of Tokyo.

The JFCR was established by Katsusaburo Yamagiwa, who is world-famous for having artificially induced a carcinoma by repeatedly rubbing coal tar into a rabbit's ear, under the slogan of "contributing to human welfare by means of cancer eradication" at the end of the Meiji Period. The existing cancer center, which has attracted attention around the world, was first organized in 1934. Initially, the Prince Hitachi was disciplined by Tomizoh Yoshida, who won the Order of Culture and is famous for identifying the Yoshida sarcoma (ascites hepatoma) during an experiment on carcinogenicity using azo dye. After the sudden death of Yoshida in

Fig. 2.17 Prince Hitachi working at the Cancer Institute around 1998 (Courtesy of the Imperial Household Agency)

1973, the Prince continued to conduct his research on fish and frog cancers, as well as comparative oncology under the guidance of Haruo Sugano and Tomoyuki Kitagawa, both of whom served as the president of the institute and were authorities in cancer pathology. He visited the Cancer Institute three times a week, to dedicate himself to his research when time with no official duties allowed. Even now, he continues to visit the Cancer Institute, which was transferred to Ariake, Tokyo, to conduct his research (Fig. 2.17).

The biological disciplines with the prefix "comparative" focus on research on the mechanism for a certain phenomenon that occurs among individual organisms. Considering that only the researches on humans and on mice closely related to them are the most important from the standpoint of medical science, it is generally believed that these comparative studies are unimportant, because they would be a kind of enumerative biology. (Incidentally, the term "enumerative biology" originally meant research that takes up individual subjects one by one. The term was used critically against, for example, a tendency to try to repeat an experiment carried out on dogs directly on cats, without digging into the essence of the problem.) That opinion is wrong. The reason is that, with the advance of genome science, it has been elucidated that the genes of lower vertebrates, invertebrates, uni-cellular organisms, etc., also play an important role in mammals, including humans, and it has become easy to analyze the entire genome of any organism other than the so-called model organisms; it is difficult to predict whether or not the

Fig. 2.18 A cancer developing in a goldfish (Courtesy of Tomoyuki Kitagawa)

knowledge obtained from the research on any organism may be useful in the treatment of human diseases, etc.

Giving an example, the findings of Osamu Shimomura, who won the Nobel Prize in Chemistry, were achieved from his research on fluorescent substances derived from the sea-fire fly (*Vargula hilgendorfii*) and a medusa, owan-kurage (*Aequoria Victoria*), based on comparative biochemistry, the usefulness of which was later proven in the elucidation of various kinds of life phenomena. It is not unlikely that the elucidation of the mechanism for carcinogenesis in lower animals may contribute to the development of therapies for human cancers. In fact, the researches conducted by the winners of the Prince Hitachi Prize for Comparative Oncology, established in 1995, include many unique works that were not well-known until they were published.

Prince Hitachi collected a variety of cancer cells, including liver cancer from lungfishes, nephroblastoma from eels and pancreatic cancer from frogs, the most especially well-known being a cutaneous cancer called "erythrophoroma" in goldfish, a color photograph of which adorned the cover of the "Journal of the National Cancer Institute, JNCI" (Fig. 2.18). Since transferring to the Cancer Institute, he has published as many as 35 papers in English, including:

"Morphologic and biochemical characterization of erythrophoromas in goldfish (*Carassius auratus*)," JNCI, Vol. 61, pp 1461–1470, 1978
"Nephroblastomas in the Japanese eel, *Anguilla japonica* Temminck and Schlegel," Cancer Research, Vol. 52, pp 2575–2579, 1992
"Polycystic kidney and renal cell carcinoma in Japanese and Chinese toad hybrids," International Journal of Cancer, Vol. 103, pp 1–4, 2003.

In almost 40% of these papers, his name is listed as the primary author. Moreover, he has published several papers in Japanese, including review articles such as:

"Tumors in mollusks: A review," Venus, Vol. 38, pp 212–214, 1979
"Thyroid neoplastic legions in fish," Igaku no Ayumi, Vol. 120, pp 1–8, 1982.

It may be considered that these papers are enough, both in quality and quantity, to qualify him for a doctorate degree in Medicine. In the entire world, there are only a few researchers who carry out researches on cancer in the same manner as he does. His research on cancer, which has continued consistently over many years, is highly esteemed throughout the world, and the German Cancer Society appointed him as its first foreign honorary member. In 2016, he received the Grand Prix of the Charter of Paris against Cancer.

Although he has never expressed any desire to acquire a doctorate in Philosophy, Chiang Mai University (Thailand), George Washington University (the US) and the University of Minnesota (the US) have all individually awarded him an honorary doctorate. He has served as the Honorary President of the JFCR since 2001, and also as the President of the Princess Takamatsu Cancer Research Fund. It's a little known fact that he is now an honorary member of the Linnean Society in London, just like Emperor Akihito.

Tomoyuki Kitagawa, who has cooperated with him in carrying out research for many years at the Cancer Institute, said, in a comment sent directly to me:

His Highness Prince Hitachi is a serious-minded researcher, who works on his research steadily. Having acquired a deep knowledge of basic biology and zoology through a series of researches over 10 years at the Department of Zoology, Faculty of Science, University of Tokyo, he took full advantage of said knowledge in his research on tumors in lower animals, centering on fish after transferring to the Cancer Institute. He is widely acquainted with many researchers and workers at zoos and fishery experimental stations across the country. He has pathologically analyzed rare tumors provided by these acquaintances, identified their characteristics (pathological diagnosis) and published papers on the findings in one after another in top-ranked journal. Being proficient in the acquisition of documents and literature as well, he gained the admiration of those around him for his frequent ability to collect literature that was difficult to obtain in those days when there was no computer-based document retrieval system, as is widely used now. All of the winners of the Prince Hitachi Prize for Comparative Oncology, which was awarded for the 17th time in 2014, have carried out unique researches. This, in turn, reflects the uniqueness of the Prince Hitachi's researches.

Prince Hitachi, a naturalist by nature, has described various new species of animal, in addition to assisting Emperor Showa with his work and conducting his own research on cancer. As for shellfish, he has co-authored several papers in both Japanese and English on Indo-Pacific file shells (Limidae and *Ctenoides lischkei*) and wentletrap shells (Epitoniidae) with Tadashige Habe, since the latter published "*Nebularia abbatis* (DILLWYN) new to Japan (Mitridae)" in "Venus," an academic journal on shellfish, in 1969.

Moreover, he co-authored a paper in English on pea crabs, which was described in the section about Emperor Showa, with Masatsune Takeda, with whom he carried out the research on the prawn *Glyphus marsupialis* (*Sympasiphaea imperialis*), described earlier, and published it in "Memories of the National Museum of Nature and Science." He also wrote several reports on wild ducks and rosy blitterling (*Rhodeus ocellatus kurumeus*).

Furthermore, in recent years, he has done research on the fauna and flora on the grounds of the Tokiwamatsu Imperial Villa, in cooperation with specialists in these fields, with the findings on birds and butterflies being reported in "Memories of the National Museum of Nature and Science" in 2005. In 2013, "Fauna and Flora at the Tokiwamatsu Imperial Villa, Tokyo," which was compiled into a report by the two men and edited by specialists at the Ibaraki Nature Museum, was published.

In 2012, he stated at a press conference celebrating his 77th birthday:

Neoplastic lesions, the benign or malignant status of which is still unknown, were identified in a sea turtle, and I am doing research on these lesions jointly with the researchers at the Cancer Institute when my official duties permit. The Tokiwamatsu Imperial Villa, my present residence, where a wide variety of birds and insects live, is said to have a great many organisms that may be observed, even though it is situated in a big city, just like the Fukiage Imperial Garden and the National Park for Nature Study, Meguro. I am conducting a joint survey on the organisms that live there with specialists in these fields.

As can be seen from this statement, he has a strong inquiring mind, even now, in spite of his advanced age. Princess Hanako has worked on the translation of several foreign fairy tales and picture books describing animals, and has donated the profits from the sales to Animal Welfare Associations, etc.

Chapter 3
Prince Akishino [1965–] and Sayako Kuroda [1969–]—The Third Generation Biologists

The Personal History of Prince Akishino

Prince Akishino Fumihito (hereafter simply referred to as Prince Akishino), the second son of Emperor Akihito, has been interested in animals since his childhood, during which he was called "Prince Aya." Various stories about his interaction with animals have been told. It was right up his alley to grasp a snake or a lizard in his bare hand, and he has the dangerous experience of having been bitten by both a raccoon dog and a dolphin. Although apparently mischievous, as may be perceived from these episodes, he was also in the habit of visiting the exhibitions on animals held at the National Museum of Nature and Science. It is said that he was the first to keep a Nagoya cochin (a local domestic fowl), a bantam and a koeyoshi (a crying cock), and then later a shamo (a fighting cock), in the Crown Prince's Palace, and speculated about the production of a new species. Moreover, he wrote about his dream of producing a new species of edible fowl 20 kg in weight, called a Bafuronze, by crossing an edible fowl, a bafukochin, with a turkey of the bronze species, in his graduation writings at Gakushuin Primary School (quoted from the book "His Highness Prince Akishino" by Emori [12] and "Three-generation Imperial Biologists" by Imajima [20]). Although he did not become engaged in research on domestic fowls until he was older, his writing shows that the genesis for walking the path of this type of research had been formed in his childhood days (Fig. 3.1).

It is said that guinea pigs, sheep, turtles, monkeys and fish, including piranha, were kept in the Palace. Furthermore, he also once kept a Japanese rat snake (*Elaphe climacophora*) in a room. Being a horseback riding enthusiast, he has participated in "Hatsunori Kai" (the first riding game of the new year) held by Ouankai, every year since his childhood, showing improvement in his riding skill each time. He had earlier become acquainted with Professor Tatsuhiko Kawashima, the father of the future Princess Kiko, through horseback riding (quoted from the book written by Akashi [1]). They say that he belonged to the Earth Science club and the Geography

© Springer Nature Singapore Pte Ltd. 2019
H. Mohri, *Imperial Biologists*, Springer Biographies,
https://doi.org/10.1007/978-981-13-6756-4_3

Fig. 3.1 Prince Aya
Fumihito (Prince Akishino)
observing a fowl kept in the
Crown Princes' Palace in
1979 (Courtesy of the
Imperial Household Agency)

and History club at Gakushuin Boy's Junior High School, and was also good at playing tennis. Moreover, he lived in New Zealand for a period of time.

Subsequently, he entered the Department of Political Studies, Faculty of Law, Gakushuin University. While a second year student, he set up a circle called the "Study Group Working on Nature & Culture," and invited Kiko Kawashima to join. As the name of the group indicates, it can be seen that he maintains an attitude of trying to harmonize two different academic disciplines, natural science and the humanities. Despite being enrolled in the Department of Humanities, he was interested in fish thanks to the influence of his father, the Emperor Akihito, especially in catfish, for reasons described later. Since his time in kindergarten, he had been taught the names of the various fish kept in a pond at the Crown Prince's Palace by his father, a specialist in ichthyology. After graduating from Gakushuin University in 1988, he studied at the Department of Zoology, St. John's College, University of Oxford, for one year, to learn about zoology in earnest. Around that time, he frequently had to return home to look after Emperor Showa, who was ill. Thus, he devoted himself intensely to his research only within a limited frame of time. When studying in the UK, he concentrated on dissection at a museum during a marathon session from about 9:00 A.M. to about 10:00 P.M. without sparing any time for rest, with the exception of a lunch break, and, in addition, he was known to sit up reading books until as late as 3:00 AM. Later, some of the findings from his research bore fruit in the form of the paper:

"Cranial anatomy and phylogeny of the South East Asia catfish genus *Belodon tichthys*" (Bulletin of the British Museum (Natural History), Vol. 57, No. 2, pp 133–160, 1991) under the names of his co-authors, G. Howes (the British Museum (Natural History)) and Fumihito, Prince Aya (University of Oxford).

When studying at the University of Oxford, he got further interested in domestic poultry, including fowls, ducks and geese, in addition to catfish, and made great effort to take pictures of 150 breeds at the genealogical facility where these poultry were being kept.

   It is generally imagined that he, the second son of the Imperial Family, has a wild and free character. On the contrary, according to the descriptions in "His Highness Prince Akishino," [12] written by Keiji Emori of the Mainichi Newspapers, who has a friendly relationship with him, he has a rather shy and stay-at-home personality and likes hobbies traditionally associated with older people. Moreover, he has not exhibited particular facility for arranging social activities in the past. Contrastingly, Crown Prince Naruhito, who seems to strike many people as being quiet and calm, has an aggressive personality, as can be seen from his preference for such activities as climbing mountains. Prince Akishino says that his personality is quite similar to that of his father, Emperor Akihito, as opposed to his appearance, which he seems to have inherited from the Shoda Family (his mother's side of the family). As a logically-minded individual, he does not immediately accept anything anyone tells him, but rather listens to the assertions of others only after he himself is well satisfied. Furthermore, he is moderate in everything. As can be understood from the above descriptions, he is most definitely descended from Emperor Showa and Emperor Akihito. Although reports were heard of his unwillingness to study subjects in his school days, he studied hard with great concentration before taking important examinations. He has continued to the present day to maintain such an attitude in regard to contemplating things through the filter of logic and practical verification since his days studying at the University of Oxford, as just described. Empress Michiko also stated, in her written response to her birthday celebration:

> He has been accustomed to considering things deeply in his heart since his childhood, and while he appears to act as if these things were no big deal, his nature of putting his all into thinking about them remains unchanged.

In 1990, he got married to Kiko Kawashima, his junior by one year at Gakushuin University, whom he had been already seeing, when the period of mourning for Emperor Showa expired and the Prince Akishino Family was created. At that time, "Princess Kiko boom" broke out into popular culture, just as "Crown Princess Mitchy boom" had before her. It was the nation's biggest celebratory event, three years before the marriage between Crown Prince Naruhito and Miss Masako Owada. In time, the Prince Akishino Family was blessed with two Imperial

Princesses, Princess Mako and Princess Kako, and, much later, with one Prince, Prince Hisahito, whose steady, healthy growth has taken place in front of the entire nation. Prince Hisahito, who is interested in all organisms and enjoys insect-collecting, is expected to become the fourth-generation Imperial biologist sometime in the future (Fig. 3.2). Although Prince Akishino's contributions to biology and other fields will be described in detail a little later, he often visits foreign countries, including Thailand, Indonesia, China, Laos and Madagascar, for fieldwork and joint researches when he has time to spare from his tight schedule of official duties. As described later, he has guided students in their studies at the Tokyo University of Agriculture, with which he has a connection, since 2001, and has also worked as a part-time lecturer and visiting professor. Since 2007, he has worked as a project researcher and special guest researcher at the University Museum, University of Tokyo. Moreover, he participated in the exhibition "Biosophia of Birds – from the Collection of the Yamashina Institute of Ornithology," which was held to celebrate the 130th anniversary of the foundation of the University of Tokyo in 2008. Furthermore, the following year, he jointly wrote and edited "Biosophia of Birds" (University of Tokyo Press) and "Summa Ornithology" (the University Museum, University of Tokyo) with Yoshiaki Nishino, the former director of the University Museum. In 2007, he also worked as a project researcher at the then Hayama Research Center for Advanced Studies, the Graduate University for Advanced Studies (SOKENDAI), from which he got his PhD, and as a guest researcher at the Center for the Promotion of Integrated Sciences, SOKENDAI during a period of 4 years starting in 2010.

Fig. 3.2 Prince Hisahito (Akishino) with a grasshopper in 2012 (Courtesy of the Imperial Household Agency)

Besides that, he serves as honorary President of the World Wide Fund of Nature (WWF), Japan, and also as President of both the Yamashina Institute of Ornithology and the Japanese Association of Zoos and Aquariums. The old Enoshima Aquarium, Kanagawa, established in 1954, was the first of a new type of aquarium, and Emperor Showa and his family visited it often. As an introduction to the respective researches on hydrozoa, gobies and catfish of the three generations of Imperial biologists, Emperor Showa, Emperor Akihito and Prince Akishino, an exhibit called "Kinjyo Heika no Gokenkyu" ("Researches of the Present Emperors") is set up at the new Enoshima Aquarium, the successor to the old one. According to the diary of Chamberlain Urabe, this room was arranged through strong appeal to those parties capable of preparing it, work that had continued under Yoshitsugu Hirosaki, the former director of the Aquarium, immediately after the demise of Emperor Showa. Hirosaki graduated from the Faculty of Fishery Sciences, Hokkaido University, and conducted his research under the guidance of Ichiro Tomiyama.

The Princess Kiko stated at a press conference held in 2002:

> While performing his official duties in good faith, the Prince continues to carry out his research, making full use of his limited time. It may have been because of the influence of the Prince that I became so greatly interested in the music and culture in the regions and countries I have visited, and that caused me to engage in the studies in these subjects with which I have become so familiar over these many years.

The above-mentioned attitude of Prince Akishino may also have been true for the Emperor Showa and Emperor Akihito; in other words, the practice of concentrating on their own specialized research within limited spare time has been passed down through three generations of the Imperial Family. The Princess Kiko took part in a club that supported people with disabilities in her days at Gakushuin Girl's High School and has kept up her interest in welfare and health-care issues. After entering Graduate School, Gakushuin University, she carried out research on the health care system for college students, making full use of the spare time from her official duties, just like her husband, and completed her master's thesis in 1995. She once again took up her research as a JSPS Research Fellowship for Young Scientists in 2009, and completed her thesis, entitled "An exploratory study on factors related to tuberculosis preventive intentions and behaviors: Results from surveys with seminar participants of the national federation of community women's organization for tuberculosis control and female college students," and received her doctorate in the humanities from Ochanomizu University in 2013. Princess Mako, the first daughter of the Prince Akishino Family, received her qualification to be a curator at International Christian University and studied museology at Graduate School, University of Leicester, for one year starting in autumn 2014. Now, she is a project researcher at the University Museum, University of Tokyo.

Achievements of Prince Akishino—Research on Catfish and the Molecular Phylogeny of Fowls

Prince Akishino has come to be known as the "Prince of Catfish." Yasuhiko Taki, a professor emeritus at the Tokyo University of Fisheries, said:

> It was at Bang Pa-In Palace, Thailand, in South-East Asia, where he visited with the members of the Study Group of Nature and Culture, Gakushuin University, that he found that he was definitively interested in catfish—.

This Palace is situated in the north of Bangkok and is surrounded by a wide moat. On one side of the moat, as many as 3–4 loaves (possibly only 1–2 loaves) of bread were sold for feeding fish. When the Prince tried to break the bread apart with his fingers, a seller requested that he give the bread to the fish without breaking it. As soon as he put the loaf of bread into the moat, according to the seller's request, a big catfish turned up on the water's surface and swallowed it in one gulp. Delighted by the spectacular scene that had just occurred in front of him, he decided to follow a path to engage in research on phylogeny and modes of life of fish belonging to the catfish family, especially the Mekong giant catfish (plaa buk, *Pangasianodon gigas*). By the way, the terms "plaa" and "buk" mean "fish" and "large," respectively (quoted from the article by Taki [59]) (Fig. 3.3).

Taki showed Prince Akishino, who was an elementary school boy at the time, pirarucu (*Arapaima gigas*) that he kept in an aquarium installed in the Institute for

Fig. 3.3 The giant catfish, *Pangasianodon gigas* (Courtesy of Yasuo Niimura)

Breeding Research (the current Research Institute of Evolutionary Biology, Tokyo University of Agriculture), at which he worked at that time. Moreover, before leaving for the University of Oxford to study, as an extra lesson, he gave guidance about ichthyological research to the Prince, who was then a humanities science student.

The Mekong giant catfish seems to be as big as 3 m in length and 250 kg in weight. Its life history and lineage were once veiled in mystery. In the 1980s, the Fishery Department, Thailand, succeeded in producing the larvae and fry of the Mekong giant catfish. Taking this opportunity, the Prince advanced his phylogenetic research in terms of morphology. The paper that he wrote under his single author name for the first time was

"Morphological comparison of the Mekong giant catfish, *Pangasianodon gigas*, with other pangasiids" (Ichthyological Research, Vol. 36, No. 1, pp 113–19, 1989).

The same issue of the journal also featured the paper "Reexamination of the status of the striped goby," written by Emperor Akihito and Katsuichi Sakamoto. Prince Akishino's research was quite orthodox and elaborate, as was that of his father, in that, for example, the paper revealed that the larvae and fry of the Mekong giant catfish appear to be morphologically similar to other pangasiids.

He wrote this paper under the guidance of Yasuhiko Taki. He expressed his gratitude to Norio Kondo, the director of the Research Institute of Evolutionary Biology and a professor emeritus at the Tokyo University of Agriculture, who was his former teacher, for permitting his use of the Insitute's facilities. Norio Kondo, who graduated from the Tokyo University of Agriculture, created seedless watermelons under the guidance of Hitoshi Kihara, a professor at Kyoto University, and established the afore-mentioned Institute for Breeding Research (the present Research Institute of Evolutionary Biology). While still a child, the Prince Akishino sometimes visited the Institute, at which Kondo, who was acquainted with Emperor Showa and Emperor Akihito, worked, affectionately calling him "Grandpa!" Kondo accompanied the Prince when he visited Indonesia and Madagascar to carry out his examination.

Recently, the Prince wrote and published, jointly with the study group of Hiroshi Kono of the Tokyo University of Marine Science and Technology, two papers on phylogenetic analysis focusing on the morphology of the larvae and fry of pangasiids (2011, 2013). According to some sources, at present, they are writing up a refined summary. It seems that DNA analysis of these specimens is also being carried out, though the findings have not yet been written into a paper (a private letter from Yasuhiko Taki). He visited Thailand several times to carry out his fieldwork for an examination of fish, including catfish, during the period from the time when he was studying at the University of Oxford to 1996. Moreover, he observed large-sized catfish living in a local area when visiting Paraguay in 2006.

By the way, I worked as the director general of National Institute for Basic Biology situated at Okazaki, Aichi, and as the president of an organization that governs three institutes, the National Institutes for, respectively, Basic Biology, Molecular Science and Physiological Sciences, for eight years starting in 2005. This type of Inter-University Research Institute Corporation, domestically and internationally valuated as top-level research institutes, has facilities and materials that are available for joint use by both foreign and domestic researchers. The Inter-University Research Institute of the Arts includes the International Research Center for Japanese Studies and the National Museum of Japanese History, while the Inter-University Research Institute of Natural Sciences includes just over 10 institutes, including the National Astronomical Observatory of Japan, the National Institute for Fusion Science and the National Institute of Informatics. These institutes have their own graduate schools, which are systematically unified into the Graduate University for Advanced Studies (SOKENDAI), situated at Hayama, Kanagawa. The professors and associate professors from the individual institutes form the SOKENDAI and guide the students in their studies. In 1996, Prince Akishino's thesis, entitled "The molecular phylogeny of junglefowls, genus *Gallus* and monophyletic origin of domestic fowls," was submitted to the meeting of the Life Science Department, SOKENDAI, by one of its members, the National Institute of Genetics. The meeting resulted in the conferment of a doctorate on him. Norio Kondo suggested that he write his thesis himself, so that he might acquire a doctorate, rather than an honorary doctorate. Among the Imperial Family members, he was the first to acquire a doctorate. There are two types of doctorate degree: a doctorate by way of advanced course and a doctorate by way of dissemination. To acquire the former, a student must master the compulsory subjects, through experiments, research and a seminar, and then submit the thesis to a board of examiners. In the latter case, a student must submit the thesis to a board of examiners through a professor, who gives guidance to the student in a special course, and only when the thesis is evaluated to be at the level equivalent to or higher than that of the doctorate by way of the advanced course is the doctorate given to the student. The Prince holds the latter type of doctorate.

His acquisition of a doctorate was reported by the mass media, bringing a level of public awareness to SOKENDAI that it had not previously enjoyed. We biologists were surprised that, given his nickname as the "Prince of catfish," he achieved his doctorate for his research on fowls. And we imagined that the general public was probably just as surprised. However, as previously mentioned, he had already shown an interest in fowls during his time at Gakushin Elementary School, and his enthusiasm for them was reignited during his period of study in the UK. As a matter of fact, during the period from 1995 to 1996, before he submitted his doctoral thesis, he worked rigorously on the three papers listed below, which were published in the prestigious international journal "Proceedings of the National Academy of Sciences, USA, PNAS":

"One subspecies of the red junglefowl (*Gallus gallus*) suffices as the matriarchic ancestor of all domestic breeds," "PNAS", Vol. 91, pp 12505–12509, 1994;
"The genetic link between the Chinese bamboo partridge (*Bambusicola thoracica*) and the chicken and junglefowls of the genus *Gallus*," "PNAS", Vol. 92, pp 11053–11056, 1995; and

"Monophyletic origin and unique dispersal patterns of domestic fowls," "PNAS", Vol. 93, pp 6792–6795, 1995.

He co-authored all of these papers, serving in the role of the primary author who assumes full responsibility for the content of a paper, and his researchers included Norio Kondo, Koji Shiraishi and Masaru Takada, all of whom worked at the Research Institute of Evolutionary Biology, and Susumu Ohno, as described in the Section "Achievements of Emperor Akihito." Takashi Gojobori, who supervised his research in the phylogenetic field, is also listed as a co-author of the third paper. Gojobori was a member of the board of directors of the Research Institute of Evolutionary Biology. Before any paper is allowed to be submitted to "PNAS," it must be recommended by a member of the National Academy of Sciences (NAS). Ohno, a member of NAS, who had promoted a series of his researches early on, recommended the Prince's papers. The contents of the papers were determined to be satisfactorily eligible for acceptance by "PNAS." The Prince observed various breeds of domestic fowl, and not only was he interested in their origins, but also wished to pursue the roots of all of the breeds of fowl during the period he spent studying in the UK. Coincidentally, having heard that there was an interesting wild fowl culture that lived and had been passed down in Indonesia, he visited the country twice to collect blood samples of both the wild fowl and domestic fowl living there. Thus, he extracted mitochondrial DNA samples from 4 species of wild

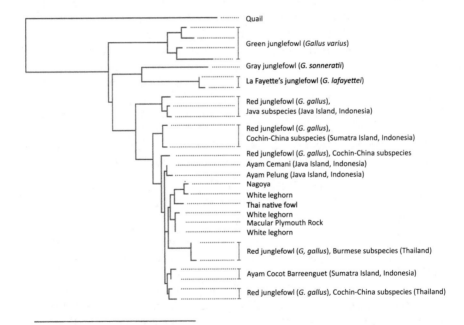

0.1 Base substitution

Fig. 3.4 The phylogenetic tree indicating the relationship between wild fowl and domestic fowl (From "Tori to Hito (Fowls and Human Beings)," [54] courtesy of Prince Akishino)

fowl and about 3 dozen domestic fowl and analyzed the base sequences in a given region with a sequencer. He went to the National Institute of Genetics situated at Mishima, Shizuoka, to analyze the data that resulted from the sequencing. Needless to say, he also joined in on the experiments conducted there. The results from the analysis revealed that, from among 4 species of wild fowl, the red junglefowl (*Gallus gallus*), the green junglefowl (*Gallus varius*), La Fayette's junglefowl (*G. lafayettei*) and the gray junglefowl (*G. sonneratii*), the most commonly kept domestic fowls might only be derived from one or two subspecies lines of a red junglefowl. Darwin's monogenesis (monophyletic theory) was supported by his analysis of red junglefowls, resulting in the settlement of the debate, which had remained unsolved for a many years after it was propounded (Fig. 3.4).

Later, it was suggested that other wild fowl might also be involved in the issue of monogenesis. In response to this suggestion, the SOKENDAI research group, including the Prince, led by Naoyuki Takahata, the then-president of SOKENDAI, used many nuclear gene (to be exact, intron) sequences, instead of mitochondrial DNA sequences, to examine any change in the genes that resulted from artificial selection in domesticating wild fowls. The resulting article, entitled

"The origin and genetic variation of domestic chickens with special reference to junglefowls *Gallus g. gallus* and *G. Varius*," was published in "PLos ONE, Vol. 5, No. 5, e10639, 2010."

It seems that domestic fowl are being kept in an environment in which they may be easily hybridized with wild fowl in the Southeast Asian countries, resulting in a mixture of genes. Moreover, the red junglefowl that is considered to be an ancestor of the domestic fowl is rich in genetic diversity, and it is estimated that human beings selected specific genes from the wild fowl living in ancient times in the process of domestication to produce a breed that has advantageous features. Prince Akishino is listed as the last author in the above-mentioned paper. He only visited SOKENDAI at Hayama a few times a year, although he communicated frequently with the members of the research group via e-mail (Fig. 3.5).

Besides this, he co-authored a total of 10 papers on molecular phylogeny and genes with other researchers, including a review article entitled

"Philological Studies on Subspecific Recognition and Distribution of Red Junglefowl" (co-authored with Takao Oka, Yoshihiro Hayashi, et al., Journal of the Yamashina Institute for Ornithology, 2004),

as well as reports on the results of gene analysis of the domestic fowl of Japan (2007) and the native fowl living in Sipsong Panna Yunnan Province, China (2008). These papers are all fully-fledged efforts.

Yoshihiro Hayashi, who graduated from the Department of Veterinary Anatomy, Faculty of Agriculture, University of Tokyo, serves as the director of the National Museum of Nature and Science. He was the first scientist since Yo Okada to be appointed as director. At present, he concurrently serves as the director general of the Yamashina Institute for Ornithology, and he was formerly the first chairman of the "Society for the Study of Human-Animal Relations." Hayashi has been giving

Fig. 3.5 Prince Akishino in discussion with his colleagues at the SOKENDAI (Courtesy of Akifumi Oikawa)

guidance to the Prince since around the time he wrote his thesis, and he still sometimes holds discussions with him that last over several hours. The Prince wrote several papers on birds, in addition to those on the humanities described later. Among them are 5 papers on the eggs of a big elephant bird (*Aepyornis*) that survived until around the 17th century. He became interested in the big bird, as well as a domestic zebu (*Bos primigenius indicus*), while visiting Madagascar. The phylogenetic works on gobies conducted jointly with the Emperor Akihito are omitted to avoid duplication. His name is listed as an originator of the study in an ichthyological paper,

"Red spotted masu salmon (*Oncorhynchus masou ishikawae*) with unique jaw morphology collected from central Japan" Ichthyological Research Vol. 56, No. 2, pp 208–209, 2009.

Taxonomy and Phylogeny

Now, let's shift our focus to two academic disciplines, taxonomy and phylogeny, to which the Imperial Family members have devoted themselves. We human beings have a native ability to recognize and classify different things and events individually. NHK (Japan's public broadcasting network) used to broadcast a popular quiz program called "Twenty Doors" (similar to "Twenty Questions") on the radio. In this program, contestants were called upon to guess about and identify things or events based on hints given to them by the MC (master of ceremonies). As I recall, they first had to select a category—"animal," "plant" or "mineral"— associated with the answer. This program assumed that we are able intuitively to judge things as organisms or non-organisms, and further as animals or plants. Similarly, we can distinguish humans from chimpanzees, despite the fact that the genomes of both primates are almost the same. Moreover, we can tell the difference between a snake and an eel, or a bee and a gadfly, or yeast and the bacterium *Escherichia coli*, either with the naked eye or with the aid of microscopes, including electron microscopes. This is applied to "minerals" as well.

In fact, the ability to distinguish and classify things and events is one of our natural endowments. More interestingly, we have a stronger sense of affinity with animals than we do with plants, vertebrates over invertebrates, mammalians over fish or amphibians, and apes over ordinary monkeys. We are endowed with a native aptitude for phylogeny. This might be because we, as humans, are also organisms.

The classification of animals and plants originated with Aristotle, along with Theophrastus, one of his disciples, in the Greek era, and approximately 500 and 520 species of, respectively, plants and animals have been recorded. In the Roman era and the Middle Ages, called "the Dark Ages," people felt no need to go any further than simply following the knowledge acquired in the Greek era. During the Age of Exploration, after the Renaissance, a variety of animals and plants began to be collected from all over the world, and a magnificent amount of knowledge regarding those specimens accumulated, resulting in the publication of books of global-scale fauna and flora. This period is called the "golden age of natural history." On the other hand, focusing on China, where herbalism had been developing from ancient days, they had not only been building up knowledge about herbal plants, but about animals as well.

In this situation, it was Linné's "Systema Naturae," the first edition of which was published in 1735, that revolutionized taxonomy and established the hierarchical classification system of animals and plants. In its 10th edition (1758), in order to describe all species, Linné adopted the binomial nomenclature in which both the "generic name" and the "specific epithet" were to be used. Thus, all organisms are now classified systematically into their corresponding kingdom, phylum, class, order, family, genus, and species. As a matter of interest, Linné's classification system of plants was comparatively artificial (artificial classification), with plants being classified into any of the following groups: those with the same number of stamen; those that are ananthous, such as mosses and ferns; and those that are thought to be asexual, such as amebae. Subsequently, as the theory of evolution became widespread, the artificial classification system became closer to the so-called natural classification, taking evolution and phylogeny into account. Nevertheless, classification was initially carried out based on apparent differences in morphology.

In this way, taxonomists engaged in fierce competition to describe new species. One barrier that they faced at the time was the issue "What is a species?", which still remains an important challenge. We may understand the reason why Darwin gave the title "On the Origin of Species" (1859) to his book on evolutionary theory. I once heard, a long time ago, that taxonomists can be divided into two types, splitters and lumpers. The former type recognizes the differences in characteristics among species in every detail and creates many new species, while the latter roughly groups organisms into the same species. Not only is morphology taken into consideration in this process, but so is development, the mode of life and other factors. Therefore, the question of what standard is to be used to identify species is the most important issue.

Ernst Meyer, an expert ornithologist and one of the recipients of the International Prize for Biology, who passed away at the age of 100 in 2005, immigrated from Germany to the US. From the 1930s to the 1950s, he promoted the General Theory of Evolution, aiming at a renaissance for natural history and taxonomy; this is neo-Darwinism based on natural selection. According to him, a "species" represents the population of organisms that can reproduce under natural conditions, and speciation into new species occurs when its reproduction becomes impossible due to geographic isolation and other factors. Yoshimaro Yamashina, also an ornithologist, reached a similar idea to Meyer's, as will be described later, but he expanded his idea to the cellular level. However, this definition has some problems; for example, it could not be applied to asexual organisms.

In the 1950s, a German entomologist named Willi Hennig proposed cladistics (cladistic taxonomy or cladistic systematics). In the past, taxonomists would intentionally pick certain characteristics in a particular way and use them as standards for classification or for understanding systems. However, this method could lead to different conclusions depending on the characteristics to be chosen. In cladistics, it is fundamental that every system of organisms be monophyletic. The term "monophyletic" indicates that the taxonomic groups (groups of species, for example) that share characteristics with each other (apomorphy) thought to have been acquired relatively recently in the process of evolution have a single common ancestor, namely, a diverging (branching) node. With the emergence of cladistics, the rules for phylogeny were set up, enabling objective discussions.

On the other hand, with the development of molecular biology and population genetics, the Neutral Theory of Evolution was presented by Motoo Kimura in the late 1960s. This theory put forward a new doctrine that most mutations that occur at the molecular levels, such as protein and DNA, are neutral to natural selection, and evolution results from the fixation of these mutations in a population by chance (genetic drift). Neo-Darwinism, in contrast, asserts that evolution results from the spread of mutations advantageous for a population's survival. Accordingly, there was a controversy between the two doctrines at that time. Today, it is known that, at both the molecular and morphological levels, there are two cases of evolution; in one case, it occurs neutrally, and in another case, it is driven by natural selection.

In taxonomy and phylogeny, characteristics (information) are used, ranging from morphological characteristics, including external form and internal structures, developmental stages, and the chromosome number and karyotype, to the electrophoretic mobility patterns of proteins such as isozymes (iso-enzymes: a group of proteins that share the same enzymatic activity), due to the advancement of bio-chemistry, and further to the amino acid sequence of a protein and the nuclear acid sequence of DNA or RNA, owing to the development of molecular biology, resulting in a remarkable expansion in the amount of available information. This opened up the way for the emergence of so-called molecular phylogenetics, in which molecular phylogenetic trees are drawn.

Giving an example, DNA is extracted from a target organism and a specific gene of interest is amplified by the PCR (polymerase chain reaction) method. The amplified product is purified and its DNA sequence, made up of four kinds of base, adenine, guanine, thymine and cytosine, is determined using a DNA sequencer. Those sequence data are analyzed by a computer to draw phylogenetic trees. Both the works of Masatoshi Nei and Joseph Felsenstein have contributed to this process as powerful tools; the former invented the neighbor-joining method and the latter applied the maximum likelihood method to the establishment of phylogenetic trees. Both were awarded the International Prize for Biology for their respective achievements. In the present day, these methods are programmed and available to everyone for drawing phylogenetic trees. Expert informaticians have also signifi-cantly contributed to this field.

In the early days, a single gene contained in mitochondria (cytochrome b, for example) was used, but the evolutionary rate varies with the gene, and a single gene does not carry enough information for the levels higher than genus or species in the taxonomical hierarchy. Therefore, a method in which multiple genes, including nuclear genes with slower evolution rates, are analyzed has commonly been used. Also, partially due to rapid advances in the performance of sequencers and com-puters, which enable human genome analysis that would have earlier required more than 10 years to be completed in just a day or less, the nucleotide sequence of more than 1,500 genes has been successfully analyzed in the most recent studies on the origin of insects. Assuming, based on the neutral theory, that the evolutionary rate of a target gene is constant (molecular clock), molecular phylogeny has enabled us to estimate the time at which genus, species, or even any of the higher taxonomical levels diverged, through comparison with geographical evidence such as conti-nental drift, orogeny (mountain-building activity), strait or channel formation and bio-fossils.

In this way, experts in taxonomy and phylogeny have acquired the ability to estimate objects and events from the past more confidently. Until now, there was no choice but to compare characteristics between living organisms. Because of this, discussions on the issues surrounding evolution are becoming active again. However, the results of molecular studies should not be overrated, and it is more important that we take steps to reveal the truth through the exchange of information from their results with that from studies in different fields such as morphology (Cf. [28, 39, 41]).

By the way, taxonomy and phylogeny, which were previously studied in uni-versities and museums, have been replaced by other fields, especially in universi-ties, not only in Japan, but also in other countries, with some exceptions. In other countries, natural history museums affiliated with universities are well-supported, and such researches are still actively carried out there. In Japan, unfortunately, even the museum founded by Edward Morse at the University of Tokyo disappeared far too early for its importance to be recognized. Even the National Museum of Nature and Science, which owns a wide variety of collections, has gradually put more emphasis on exhibitions. Such a situation is not very different from that at other museums in Japan.

Above all, it was a serious blow to the development of taxonomy and phylogeny that many laboratories working in these fields have been closed at the major universities, including the University of Tokyo. Still, at other universities, a relatively large number of researchers in these fields have delivered lectures on their basic knowledge to the students in the Department of General Education. Nevertheless, even such departments were eventually closed at many universities as a result of the strong demand for "specialized education in earlier stages of students' careers," which was made some years ago, leading to the closure of avenues for educating their successors.

Koscak Maruyama, one of my university classmates, who was the president of Chiba University and afterwards served as the director of the National Center for University Entrance Examinations, attempted, some years ago, to gather experts in taxonomy and phylogeny who were working individually all over Japan so as to launch an united graduate school. Unfortunately, his attempt did not come to fruition. In truth, he took this initiative to fulfill such a dream aspired to, but never attained, by Yo Okada, who worked as the director of the National Museum of Nature and Science in the 1950s, that seven animal and plant laboratories would be established in the museum. This conceptual plan was barely realized in the form of a graduate course solely between the University of Tokyo and the museum, under the prodding of Tsuneaki Kawamura, the director of the museum at that time (Maruyama [38]). I believe that, ideally, the museum should establish its own graduate school and form an alliance of researchers in the fields related to evolution so as to educate their successors and organize collaborations among them.

The National Museum of Nature and Science and the universities are administrated by two distinct bureaus in the Ministry of Education, Culture, Sports, Science and Technology (MEXT): the Lifelong Learning Policy Bureau and the Higher Education Bureau, respectively. Although this makes the problem more complicated, I sincerely hope that the plan will soon be put into motion, as it is never too late to start. Nevertheless, only recently have university museums (university-affiliated museums) been established or renovated at the former imperial universities and other institutions so that researches in this field can take place in an improved environment, even if just for a bit. It provides somewhat good news for us that the number of research laboratories has been slightly increased, combined with the tendency for the number of young researchers in the field to increase as well with the foundation and development of the Society of Evolutionary Studies and others in Japan.

Biostory (Biosophia Studies)

Going back to the previous topic, Prince Akishino has been interested in a harmonization of the natural sciences and the humanities, as described earlier. For convenience, we set aside one section in regard to this subject separate from that which describes his biological works, but in his mind, he has long carried the idea that both the natural and human sciences should be harmonized. In 1989, when his first paper on catfish (plaa buk) was published, he gave a presentation with Yasuhiko Taki entitled "Pangasiids as a promising aquaculture resource in Southeast Asia—Biological characterization and genetic relationships," in front of audiences attending a meeting of the Asian/Oceanian Society of Aquaculture. His contention was to clarify the overall picture for this big catfish, which is almost unknown in terms of biology and fork culture. In 1996, he visited Thailand with Osamu Akagi of the Osaka University of Foreign Studies, Tomoya Akimichi of the National Museum of Ethnology (who was later transferred to the Research Institute for Humanity and Nature), and others, to make an examination of the mode of life and habitats of plaa buk, as well as the methods, practices and traditions involved in fishing. The results from the examination were compiled into a report:

"Ethnic-Ichthyological Study of Plaa Buk (*Pangasianodon gigas*) in Amphoe Chiang Khong situated in the northern area of Thailand," Bulletin of the National Museum of Ethnology, Vol. 21, No. 2, pp 293–344, 1996.

He writes, in the book entitled "Namazu – imeiji to sono sugao" ("Catfish: Its Image and Real Mode of Life") (Yasaka Shobo, 2008), co-authored with Tomoya Akimichi, Hiroya Kawanabe, the director of Lake Biwa Museum, et al.:

> Since Linné wrote "Systema Naturae," the nomenclatures consisting of "genus" and "species" have been given to a wide variety of organisms based on their appearances. In short, this naming method is binominal nomenclature, and the designations given using this method are called scientific names. This has enabled us to recognize organisms by means of their commonly used names in spite of differences in culture. It can be said that this practice is a prominent fruit of contemporary natural science.

> On the other hand, a question, "Is it right to interpret organisms in accordance with such a nomenclature system only?," has arisen. This question has still not been settled. For example, we human beings always tend to recognize any person standing in front of us by means of their personal names, as well as features and positions, rather than as being *Homo sapiens*. I think this is equally true in our contact with other living things.

Moreover, he published many photographs of fowls taken during the period that he studied in the UK, along with explanations of them, as well as of the European culture and history relating to poultry in "Illustrated Encyclopedia of European Poultry" (Heibonsya, 1994), already published after his return to Japan. Although DNA-based phylogenetical researches have allowed us to estimate the species of wild fowl (ancestors) from which domestic fowl had been derived and the place where domestication had been originated, the incentive for domestication, i.e., "why wild fowl had been domesticated," and the further process through which a wider variety of breeds have been produced have not yet been revealed.

He writes in his book "Tori to Hito – Minzoku Seibutsugaku no Shiten Kara" ("Fowls and Human Beings – from the Ethno - Biological Viewpoint") [54]:

> It is very important for us to elucidate the reason why human beings kept domesticated wild fowl and the ideas based upon which they have produced various breeds of fowl. In other words, I have always kept it in mind that, before I can understand domestication, I have to be aware of the fact that domestication is a culture in and of itself.

It is said that red jungle fowls, which often fight with each other, have the highest-ranked volcanic temper among birds in the family Phasianidae. The Prince has pointed out that, in the countries in Southeast Asia, cockfighting is popular; accordingly, cockfighting events were associated with animal fortune-telling and rituals, leading to domestication. He has carried out his researches on the development of a free-range fowl tracking system and on the morphological function of the musculoskeletal system of a fighting cock from the above viewpoint. In addition, he wrote a paper entitled:

"Applying the Kansei model to the preference of chicken figures in Japan and Thailand - Based on an investigation of a preference for chicken figures in Chiang Rai and Okinawa," Journal of the Yamashina Institute for Ornithology Vol. 40, No. 1, pp 23–42, 2008.

Furthermore, he edited, with Princess Sirindhorn of Thailand, a book in English entitled

"Chickens and Humans in Thailand: Their Multiple Relationships and Domestication," Siam Society,

which features the results from a Japan-Thailand joint research that started several years ago. As can be seen from the above-mentioned joint research alone, a close relationship has been established between Japan and Thailand. Recently, he also visited Thailand, Laos and China, where domestication seems to have originated, to conduct an examination of fowls.

In 2003, he also participated in the establishment of the Society of Biosophia Studies, together with other researchers who shared his intention to develop a form of multidimensional recognition of organisms. He gave his presentation at the Society's regular meeting and, at present, serves as an executive director of the Society. The organisms targeted in the Society include imaginary creatures such as the kappa (river imp) and the ogre. He originally preferred the term "creatures" to the term "organisms." He has co-edited the following papers with his fellow researchers:

"Study of Human-Animal Relations, Vol. 1, Human View and Representation of Animals." Iwanami Bookstore, 2009.
"Study of Human-Animal Relations, Vol. 2, Livestock Culture." Iwanami Bookstore, 2009.

"Encyclopedia of Catfish." Seibundo-shinkosha, 2016.

He wrote about 40 papers or more, though this figure may be inaccurate, because papers evaluated as being an achievement in the field of the humanities, including books, reviews and essays, are a bit different from those in the field of natural science.

He has engaged in a wide range of activities, including serving as an editor of a book:

"Livestock and Poultry in Japan (Field Best Encyclopedia, Special issue)," GAKKEN, 2009.

Naturally, he is also interested in the issue of biodiversity and is committed to the conservation of valuable breeds of fowl produced by human beings.

Takashi Gojobori, his thesis supervisor, has stated:

> Prince Akishino has a rich knowledge of biology. I am guessing that the background to this was the tutelage of the Emperor, his father. He had worked so hard for half a year before completing his doctorate thesis that he sent his manuscript to me by FAX late at night. His very positive attitude toward making his study interdisciplinary, going beyond the boundaries between academic disciplines such as biology and cultural anthropology, is admirable. In future, I recommend that he further narrow his theme so as to carry out even deeper research. He is a researcher of top-level logicality and knowledge.

Tomoya Akimichi also admires him, saying, "His research covers a wide range and extends to a profound level free from the existing academic disciplines by finding harmonization between different fields, such as biology, archaeology and folklore" (quoted from a book written by Emori [12]).

Yoshihiro Hayashi, who has kept a close eye on Prince Akishino's works, provided me with a comment in a private letter:

> It is familiar to a lot of people that Prince Akishino has continued his interest in catfish. However, he acquired his doctorate for research on fowls, in which he has been interested since he was a child. At present, he is focusing more strongly on "chickens" as a cultural creature, rather than "fowls" in terms of biology, especially in regard to the domestication and development of new breeds, as well as on the background against which a wide variety of breeds have been produced throughout the world. Actually, every time he visits local areas around Asia, he makes a habit of talking with the people of the villages to discuss the domestication of wild fowl and the development of new breeds in terms of their recognition of fowls, especially as relates to preference of feather and leg colors and cultural taboos associated with them, as well as rituals and fortune-telling, which might induce the production of new breeds. Specifically, not only has he been strongly interested in the multidimensional relationships between human beings and animals, but he has also extended his interest to the modification of all wild birds and animals into poultry and livestock, in addition to fowls. What should be noted is that he throws himself into the role of a coordinator of a research project, rather than just a researcher.

Respective honorary doctorates were conferred upon him for his contribution to researches on catfish and fowls by universities that he has associated with in Thailand, including Kasetsart University (biology of fisheries) (1995); Chulalongkorn University (science) (2001); Ubon Ratchathani University (agriculture) (2003); King Mongkut's Institute of Technology (science of fisheries) (2007), and Mahasarakham University (biology) (2018).

The International Biology Olympiad and Prince Akishino

The International Science Olympiad (ISO) is a global-scale science contest for high school students. In the contest, high school students compete with each other in their knowledge of subjects of science; it began with mathematics (1959), with physics, chemistry, information, biology, astronomy, geography and geoscience, in chronological order, being subsequently added. Belatedly, Japan has participated in the categories of the Olympiad, including mathematics (1990), information, chemistry, biology, physics, geography and geoscience, again in chronological order. It has been reported by the press and other media that the Japanese students were awarded the gold prize in the categories in which Japan participated, mathematics and information, at a relatively early stage. Japan participated in the biology category (IBO) in 2005. As with other subjects, IBO was established to awaken high school students' interest in and attention to biology throughout the world: developing their abilities; growing human resources capable of contributing to the prosperity of their countries in the future; providing an opportunity for the exchange of information for the benefit of biological education in countries throughout the world; and providing a valuable opportunity for international exchange among the participants.

ISO takes place over the course of a week, during which it gives experimental and theoretical questions to the participants. Between examinations, an excursion is carried out so that the participants can learn about the different histories and cultures of their individual countries and expand the wave of exchange. On the last day, the gold medal is given to 10% of the participants, the silver medal to 20% and the bronze medal to 30%. Although this may make it seem like it is easy to acquire one of these medals, it is actually rather difficult, considering that promising youth from all over the world gather at the test center all at once to try to win the medals.

Upon my sudden appointment as the chairperson of the Japan Biology Olympiad (JBO) Committee in February 2005, I hastily supervised the internal preliminary round during a period of only two to three months and sent four students off to Beijing, where the final round would be held. The Japanese participants won two bronze medals and ranked 31st among 50 participating countries in by-country ranking. China, the US, Thailand, Korea, Taiwan and Singapore stood high, as always, and Japan was lowest-ranked among the so-called advanced countries. In the following year, Japan did not improve much, only earning three bronze medals. It was then clarified that a thick textbook, "Campbell's Biology," which is used in the general, or basic, educational course at Japanese universities, had been used to set a range of possible questions. For this reason, we made the selected high school students read the book. This gradually led to a good result, and Japan subsequently won higher-ranked medals: three silver medals and one bronze medal, compared with the previous result of one silver and three bronze. Meanwhile, despite the fact that Japan had just participated in the Olympiad, a challenge was abruptly issued to the Japanese Committee to hold the 20th International Biology Olympiad in 2009, a source of much confusion for us. The reason why we were confused was that the

International Chemistry Olympiad Committee had already decided that the Olympiad would be held in Japan in 2010, and we had to solicit donations on behalf of both Olympiads from the same industries and communities. Fortunately, we achieved our target amount of donations directly before the "Lehman Shock" that occurred in the year before the Olympiad was to be held, thanks to the efforts of Hiroo Imura, the former president of Kyoto University and a member of the Council for Science and Technology Policy, and Katsunosuke Maeda, the Chairperson of the Board of Directors, TOREY Industries, who had previously been devoted to the promotion of education and research into natural sciences, so that successfully settled that problem. A large amount of funding was also provided by the Ministry of Education, Culture, Sports, Science and Technology (MEXT), and the Olympiad was successfully held at the University of Tsukuba.

The Committee decided to ask Prince Akishino to serve as the honorary president of the Olympiad, and I myself visited the Prince's Palace to deliver the request on behalf of Hiroo Imura, the Chairperson of the Olympiad, with Osamu Numata of University of Tsukuba, who was the chairperson of the executive committee. The Prince accepted our request in a friendly manner, and we felt greatly honored when he arose from his seat to see us out with Princess Kiko.

Partially because the International Olympiad was being held in Japan, the number of participants in the internal preliminary round considerably increased to more than 2,000 (at present, it is a little less than 4,000) from about 300 at the beginning. The Olympiad was successfully held in the presence of Prince and Princess Akishino in July, and the Prince, owing to his strong interest in biology, presented a message that deeply impressed the audiences (Fig. 3.6). Japan's one gold and three silver medals resulted in the nation ranking 6th in by-country ranking. Furthermore, at the Olympiad held in Taiwan two years later, Japan achieved even more satisfactory results, with its three gold and one silver medal causing it to be ranked 3rd.

The winners of ISO medals have traditionally been invited to a special celebration by the Prime Minister and the Minister of Education, Culture, Sports, Science and Technology since Junichiro Koizumu served as a Prime Minister of Japan, as in the case of winners at the Olympic Games. The number of universities that matriculate the winners of ISO medals unconditionally has gradually increased. In fact, the students picked out for ISO are blessed with natural abilities. Some of the statistics show that many winners of Fields Medal worldwide have come out of the ranks of the winners of ISO gold medals. On the other hand, I have heard that some high schools refrain from participation in the internal preliminary round of the ISO for the reason that the students, if they participate, may not have enough time to prepare for the entrance examination. I would like to place the issue before the Japanese educational establishment that, generally speaking, secondary education should be designed and implemented so that the unique personalities and talents of the individual students can grow and come to fruition. However, in order to realize such an ideal education, we will need our teachers to make their best efforts. Moreover, for the teachers to get good results, sufficient time is needed. The

Fig. 3.6 Prince Akishino delivering his speech at the International Biology Olympiad (Courtesy of Osamu Numata)

"Course of Study" defined by MEXT should state the learning contents to be mastered at a minimum level, but not at the maximum level.

At the press conference in 2009, Prince Akishino stated:

> I'm glad that Ryota Ohtsuki, one of the Japanese participants, won the gold medal at the IBO held in Japan this year, in which I was involved as the honorary president. This was Japan's first notable feat since our first participation in 2005. The remaining three participants also achieved excellent results, all winning silver medals. By the way, if my memory serves me, all of our participants won medals at the three consecutive IBOs. I'm glad to know that there are students who are eager to learn biology, as mentioned above.

Sayako Kuroda and the Yamashina Institute for Ornithology

Princess Nori Sayako, the first daughter of Emperor Akihito and Empress Michiko, was educated under the idea that she would follow her destiny and leave the Imperial Family someday, according to the existing Imperial Household Law. For this reason, the Empress often went on short trips around the country together with her daughter, as is well known. She was devoted to Japanese literature and Japanese dancing, and was known to have shown affection to animals in general since she was a child. It is said that she gained particular pleasure from observing birds, starting when she was a high school student. While majoring in Japanese literature at the Department of Japanese Literature (current Department of Japanese Language and Literature), Faculty of Literature, Gakushuin University, she started up a bird-watching circle with her friends, inviting a researcher from the Yamashina Institute for Ornithology to serve as a teacher and visiting the Institute to aid in the researchers' works. Later, not surprisingly, she would go on to work at the Institute as described below. She also likes dogs and has participated willingly in activities related to guide dogs since she was a student at Gakushuin. After graduating from the university in 1992, she worked at the Yamashina Institute for Ornithology situated near Teganuma (Tega Pond) in Abiko, Chiba, as a part-time research assistant, at her own request, and then as a part-time researcher. By the time she resigned her duties to get married, she had been organizing literature and materials there for more than ten years. She was the first Princess to earn money by working. Meanwhile, she did research on birds on the grounds of the Imperial Palace and the Akasaka Imperial property, taking particular interest in a kingfisher, which she herself tagged with a label on its leg. She also placed a vertical blind in front of its nest so that she could patiently observe its comings and goings, carrying out a study of its mode of life (Fig. 3.7). A waka poem that she wrote in 1996 well reflects the feeling of pleasure this activity afforded her:

> At the crack of dawn in summer, fledglings depart to the sky one after another, implying that the time has come for them to leave the nest (quoted from the book "Hi-to-Hi wo Kasanete (Days Went by)" by Princess Nori Sayako [56]).

The results of her research on kingfishers were published in the "Journal of the Yamashina Institute for Ornithology," as the joint works as listed below:

"Breeding of kingfishers on the Akasaka Imperial property," Vol. 23, pp 1–5, 1991; and
"Study of the breeding of kingfishers (*Alcedo atthis*) at the Imperial Palace and Akasaka Imperial property," Vol. 34, pp 112–125, 2002.

In addition, she co-authored papers with other researchers, publishing the results of her general observation of birds in "Memoirs of the National Science Museum" as listed below:

"Avifauna in the Imperial Palace, during the period from April 1996 to March 2000," Vol. 35, pp 7–28, 2000; and
"Avifauna on the Akasaka Imperial property during the period from April 2002 to March 2004," Vol. 39, pp 13–20, 2005.

Moreover, she has contributed articles to journals about birds and wrote the entry on kingfishers in the "Birds" section of "The Encyclopaedia of Animals in Japan," published by Heibonsya. Immediately before her marriage, she edited and published the "Catalogue of Birds in John Gould's Folio Bird Books," under the name of Princess Nori-no-miya Sayako (Tamagawa University Press, 2005).

According to the "Afterword" that she wrote, she was responsible for organizing a total of 23 volumes of John Gould's Folio Bird Books, which was collected by

Fig. 3.7 Sayako Kuroda, Princess Nori at that time, watching wild birds in 1992 (Courtesy of the Imperial Household Agency)

Nobusuke Takatsukasa while studying in the UK, into an album at the Yamashina Institute for Ornithology, replacing the old scientific and English names of birds given to them more than 100 years ago with new ones. However, deeply impressed by the books, Testuro Obara, the honorary president of Tamagawa University, collected the volumes independently in 1999 and found that they contained illustrations that were either not in the possession of the Yamashina Institute for Ornithology or were versions that were different from those at the Institute. For this reason, she conducted an examination of the organization of the Books, with permission from the university. The secure, several-year-long examination covered almost all of the illustrations. This work was originally intended to prepare materials for the internal use of the Institute, and the complete list of illustrations came to be published partially because of a recommendation by the university. Ryozo Kakizawa, the former vice president of the Institute, who served as a professor at Tamagawa University, reviewed the "Catalogue of the Birds in John Gould's Folio Bird Books" as follows:

> In the case of some species, it is often difficult to determine the existing scientific names from the figures and the scientific names of the birds in the original illustrations through

analogical reasoning. This study was only intended to trace the history of the scientific names of birds. … This catalogue must serve as a milestone for researchers who are going to carry out research on "John Gould's Bird Books" and be so useful to them, especially to those who are biographically or artistically interested in the Books, that they cannot part with them. [32]

Princess Nori Sayako married Yoshiki Kuroda, one of her brother Prince Akishino's classmates at Gakushuin and a Tokyo Metropolitan Government official, in 2005, and was renamed Sayako Kuroda, thereby officially leaving the Imperial Family. Although the Yamashina Institute for Ornithology eagerly asked her to continue her work at the Institute after her marriage, her official departure from the Imperial Family forced her to decline all such requests. Similarly, she refrained from all of the official domestic and foreign duties that she had performed faithfully up to that point. However, she has worked at the Museum of Education, Tamagawa University, as a visiting research fellow since 2008 and served as a guide when the Emperor and Empress, and Prince and Princess Akishino, visited the exhibition of Gould's birds held at the university. It is pleasing to learn that she has come back to the Yamashina Institute for Ornithology as a part-time researcher. Makoto Watanabe describes her as such:

> She has an attentive, simple, reliable, common-sense, clever and straightforward personality … in short, despite being small in stature, she is an admirable lady, who always does her best in everything. [68]

She performed her respective duties faithfully while a Princess. In my interaction with her, I have also gotten the impression that she is an intelligent but reserved lady. Similarly, Princess Noriko, the second daughter of the late Prince Takamado, whose wedding has recently been arranged to Kunimaro Senge, the first son of Izumo Taisya Guji (a Shinto priest), enjoys bird-watching. It is said that Princess Takamado Hisako, her mother, who studied anthropology and archaeology at the University of Cambridge, would often take them around to visit the places where bird-watching activities took place.

As described earlier, the Yamashina Institute for Ornithology was founded by Yoshimaro Yamashina (1900–1989), a cousin of Emperor Showa, as the Yamashina Bird Museum on the Yamashina grounds situated at Nanpeidai, Shibuya, Tokyo in 1932. The vast site of the Yamashina grounds at Kojimachi, Tokyo, where Yoshimaro was born, has an area as large as 16,500 cm^2. The Prince Yamashina Kikumaro, his father, not only showed interest in the observation of weather and seismic activity, but also preserved many specimens of bird that he caught on the grounds. It is said that the first specimen in the Yamashina Collection was a pair of stuffed mandarin ducks, which Yoshimaro, who had been interested in birds from his toddler days, had asked his father for and finally got at 6 years old. He grew into a military man, following the custom of the Imperial Household, and became a commoner when he was 20 years old, establishing the marquis Yamashina Family, owing to his status as the second son. Meanwhile, his interest in birds having only gotten stronger, he finally resigned his position and entered the University of Tokyo as a special student in 1929 to study zoology for two years.

The Institute's building was structurally secure enough for the specimens to survive fire, even when it was directly hit by incendiary bombs dropped by the US Forces during the war.

The species and modes of life of birds living in Japan had been generally clarified by the European and US researchers, including Blakiston, since Siebold, until around the middle of the Meiji Period. The names of Japanese ornithologists such as Isao Iijima, Nagamichi Kuroda and Seinosuke Uchida would not been known until, at the earliest, the period from the end of the Meiji Period to the Taisho Period. Against this background, for its initial 10 years, Yamashina energetically collected the specimens of birds that could be found on minor islands such as Ogasawara, Chishima (Kuril Islands) and Ryukyu, as well as those on the Korean Peninsula and Taiwan, including Sakhalin, Manchuria and Micronesia, which had just recently come under Japanese rule at that time. Moreover, he did research on the modes of life of the birds and bred them. The results were collected in a book, "A Natural History of Japanese Birds," Vol. 1 (1933–1934) and Vol. 2 (1941). The contents of the books are esteemed as being extremely valuable. He had a plan to publish successive Volumes 3 to 5, but was forced to give them up because of the outbreak of the war. In the process of describing new species and subspecies of birds, Yamashina encountered the question, "What is a species?"

Cross-breeding between a male donkey and a female horse produces a mule suitable for physical work but incapable of reproduction, which is called the first filial generation (F1). If such a cross-breeding allows F1 and later generations of birds capable of reproduction to be produced, they are assumed to be of the same species, in accordance with the definition advocated by Ernst Mayr, et al. Having doubts about the classification of species based only on morphology, Yamashina tried to elucidate this kind of issue in terms of cytology. He studied under Mamoru Oguma of Hokkaido University and observed the process of the production of sperm cells through cross-breeding between various species. From the result, he recognized that any difference between species and infertility associated with crossbreeds were attributable to non-homology in a chromosome, such as its number and shape between species, and applied this type of classification to Order Anseriformes (geese and ducks). The research results were compiled into a book, "Classification of Animals Based on Cytology" (Hoppou Shuppan Sha, 1949), which was awarded the Prize of the Genetic Society of Japan the following year [71]. I remember reading this book and being impressed by it in my student days. His work was pioneering because the genetic-biochemical classification of birds had not been carried out yet, and would not be until 1970. However, since it was in the middle of a chaotic post-war period, the peerage system collapsed, leading to financial difficulty at the Institute. Yamashina decided to give up continuation of his own research and sold his pieces of land one after another without hesitation, so as to prevent the Institute from having to part with valuable specimens and books. He served as the president of the Ornithology Society of Japan and the 2nd president of the Association for the Preservation of Birds (the current Japanese Society for the Preservation of Birds).

The current Yamashina Institute for Ornithology (YIO), which was moved to Abiko, has almost 70,000 specimens of bird archived, including extinct species from all over the world, houses almost 40,000 books, including a number of quite old ones, and plays a role as a core center for international ornithological research. As described earlier, the President and the Chairman of the Board of Directors of YIO are, respectively, Prince Akishino and Hisanaga Shimazu, the husband of the former Princess Suga Takako, suggesting its deep relationship with the Imperial Family. The Institute has been endeavoring to collect specimens and books about birds and integrate the collection into a database, while joining the study of migration behavior and the lifespans of birds based on individual identification using leg bands, as well as the DNA-based study of all birds. It is carrying out a study on the conservation of extinct species such as albatrosses and Yanbaru rails (*Gallirallus okinawae*) and launched a project to develop new breeding grounds for albatrosses using decoys (models) in the year when Princess Sayako entered the Institute, having gradually achieved success.

Crown Prince Naruhito's Research

Finally, I would like briefly to introduce the achievements of Crown Prince Naruhito, who is also an academic person. The two-generation Emperors, Showa and Akihito, showed interest in biology and history, and Crown Prince Naruhito, who majored in history at the Department of History, Faculty of Literature, Gakushuin University, selected to pursue the latter discipline. However, according to the book written by Minoru Hamao, who served three Imperial Family members, Emperor Akihito, the Crown Prince and Prince Akishino, [15] the Crown Prince collected insects around Karuizawa in his early childhood. It is said that he treated insects as if they were his friends. Since that time, he has gone mountain-climbing. He is also good at horseback riding and keeps dogs and cats in the Crown Prince's Palace. The fact that he selected the Faculty of Literature rather than the Faculty of Law or the Faculty of Economy probably reflects the strong will of the Imperial Couple, his father and mother, who hoped that he would follow the path to which he aspires.

He writes, "I have been very interested in "roads," which are vehicles for transportation, since I was a child," and "I was impressed by "Oku no Hosomichi" ("Narrow Road to Oku", written by an eminent haiku poet, Basho Matsuo), which I read with my mother when I was a school boy at Gakushuin Primary School" in "Along with the Themes: Two Years During Which I Studied in England," an account of his time studying in the UK [10]. Maybe it is his interest in roads that has caused him to so enjoy jogging and mountain-climbing. He joined the Earth Science Club and the Geography and History Club at Gakushuin Junior High School. At Gakushuin High School, he acquired a wide range of knowledge about the successive emperors in Japan, "Kojiki," "Nihonshoki," "Manyoshu," cultural history, current affairs, etc., as part of the studies required for an Emperor, through the delivery of lectures by teachers. For his graduation thesis, "A Consideration of Marine Transport in the Inland Sea in the Medieval Period," he selected the history of transport and communications in Japan's Medieval Period, especially marine traffic, as his research theme, and did research on the actual conditions for commodity distribution, focusing on salt, rice and timbers, in the Inland Sea.

He entered the Master's Programs, Graduate School of the Humanities, in 1982, and further studied at the Graduate School, University of Oxford, for more than two years from June 1983 to October 1985, receiving the guidance of Peter Mathias, a specialist in the modern economic history of the UK. Akira Fuji, a chamberlain to the Crown Prince, lived near the dormitory to serve him as a counselor. He selected the history of water transport along the Themes in the 18th century as his research theme. At that time, coal was carried upstream from Newcastle via London along the Thames, flowing through the center of Oxford Town, while malts essential for brewing beer were transported downstream. A conflict between, among others, water transportation businesses that used ships and millers who used a water mill remained unsettled, and a lock gate was built between the high and low water levels. He revealed the above facts through deep research into primary materials from those days and compiled them in

"The Thames as a Highway" (Oxford, 1989).

As seen in other members of the Imperial Family, his efforts in the name of carrying out his works are steady and elaborate. In this way, he fully enjoyed his first and last taste of free student life, acquiring a wide range of knowledge of the worldly circumstances and the situations in foreign countries, both in pubs and from the public. He got his MA in the humanities in 1988, after returning to Japan. Since 1992, he has worked at the Archives, Gakushuin University, as a visiting researcher, so that he might continue his research on the medieval history of Japan, and delivers lectures at Gakushuin University and Gakushuin Women's College. The University of Cambridge granted him an honorary doctor of law degree. It should be noted that, unlike the doctorate of natural science, the doctorate of the humanities or the social sciences is generally difficult and requires time to earn, though such conditions have been considerably improved.

In the following year, 1993, he entered into a happy marriage with Masako Owada, a diplomat, and in a reflection of the wedding celebration parade held for the Emperor and the Empress more than 30 years earlier, the nation blessed them with joyous expression. Although it took time, the couple was blessed with a child, Princess Yoshi Aiko, who grew up to be a student at Gakushuin High School.

Prince Naruhito has participated in many domestic and international events and was appointed as an honorary president of the "United Nations Secretary-General's Advisory Board on Water and Sanitation (UNSGAB)" in 2007. He is the first Imperial Family member to assume an official position at the United Nations. Taking advantage of his expertise, he has presented lectures entitled "Waterways Connecting Kyoto and Local Regions" and "Edo and Water Transport" at the 3rd and 4th World Water Forums, respectively, held by the Advisory Board, at which researchers on this theme got together from around the world. In these lectures, he stated the results of his recent research on water transport in Japan while mentioning the results of his study at the University of Oxford. He presented the keynote address at the UN Special Thematic Session on Water and Disasters held at the UN Headquarters in 2013, as a result of which the academic world acknowledged his superior expertise. I have heard indirectly that he has carried out his research in the field of the science of the humanities completely by himself, unlike in the fields of natural science and biology, in which joint researches are usually conducted, and a vast number of books on water are archived in the Crown Prince's Palace. According to the recording of a message that Crown Princess Masako released for a press conference, he has a deep knowledge of plants, and, together with the Crown Princess, had long kept a stag beetle that he found at the Palace as a pet. This is a graphic illustration of the way in which all of the Imperial Family members have communed with living organisms.

Chapter 4
The International Prize for Biology

The International Prize for Biology—Aiming to Establish a Prestigious Award in Biology Equal to That of the Nobel Prize

As is well known, the Nobel Prize was established in accordance with the last will of the Swede Alfred Nobel, the inventor of dynamite, in 1901. Since then, five Nobel Prizes, the Nobel Prize in Physiology or Medicine, the Nobel Prize in Physics, the Nobel Prize in Chemistry, the Nobel Peace Prize and the Nobel Prize in Literature, have been awarded every year. In 1969, the Sveriges Riksbank Prize in Economic Sciences in Memory of Alfred Nobel (the Nobel Prize in Economic Sciences) was added to these prizes. Japan has produced 23 Nobel Prize winners so far: 9 in Physics, 7 in Chemistry, 4 (plus 1 in 2018) in Physiology or Medicine, 2 in Literature and 1 in Peace, which is the highest number among non-Western countries. It should be noted that, in addition to above-listed winners, Yoichiro Nanbu and Shuji Nakamura won the Nobel Prize in Physics. Both winners had worked on the research subjects for which they received their prize in Japan, but were US citizens when they won it. Although it is the rule, stemming from the instructions left by Nobel himself in his will, that the nationality of the nominees for the Nobel Prize are not to be considered, the number of Japanese winners ranks 6th among countries in the world. Generally, the number of winners from non-Western countries is very small compared with that of Western countries. Incidentally, the number of US winners has reached about 340 or more, among a total of roughly 920 winners by 2017. Among the Prizes in the natural sciences, the winners of the Nobel Prize in Physiology or Medicine are selected by the Nobel Committee formed at the Karolinska Institute in Stockholm. It is believed that the winners are named by the Swedish members of the Committee; however, the Committee's rules state that the members of the Committee, the selection process and all other information shall not be disclosed for a period of 50 years. The first Nobel Prize in Physiology or Medicine was awarded solely to Emil Adolf von Behring, a German

© Springer Nature Singapore Pte Ltd. 2019
H. Mohri, *Imperial Biologists*, Springer Biographies,
https://doi.org/10.1007/978-981-13-6756-4_4

doctor, but not to Shibasaburo Kitasato, von Behring's coworker, who conducted the research on "serum therapy for diphtheria." This was merely the beginning of a long period during which Japan was given no opportunity to win the prize, lasting up until the time when Susumu Tonegawa became the first Japanese researcher to receive the Nobel Prize in Physiology or Medicine, for his discovery of "the genetic principle for the generation of antibody diversity," in 1987. In the meantime, the prize was given to Johannes Andreas Grib Fibiger, a Danish pathologist, who had developed cancer on *C. elegans*, but not to Katsusaburo Yamagiwa, who had produced the world's first artificial cancer using chemical substances in 1926. However, it has since been demonstrated that the results of Fibiger's research were not valid. Although the prize was sometimes given to more than one contributor to a professional field at that time, which did not happen in the case in which Kitasato was denied the prize, it is very regrettable that this kind of situation happened. The prize would have been given to Umetaro Suzuki for his discovery and crystallization of Vitamin B1 (he named it Orizanin) if it were to happen today, as seen in the case in which Koichi Tanaka's Japanese presentation of his research alone took priority over those of other nominees, resulting in the Nobel Prize in Chemistry being awarded to him in 2002. Moreover, another example is the case of Jokichi Takamine, who discovered adrenaline. Thus, the number of winners of the Nobel Prize in Physics, from Hideki Yukawa, the first winner (1949), to Shinichiro Tomonaga, and then Reona Ezaki, opened up a big lead over that of the Nobel Prize in Physiology or Medicine, partially because of being bad luck.

Against this background, after World War II, the greatest importance was attached to the material sciences, focusing on physics in Japan's natural science community, and a far lesser chunk of the budget was allocated to researches in the life sciences. Regarding laboratory space as well, only small or average-sized areas were assigned for biology compared with those for physics and chemistry, for the reason that researchers in the life sciences might only need a microscope to do their research. Such unfair budget allocation tends to occur when the pie from the budget source is so small. Many years later, I conducted a survey of the budgets allocated to all of the national and public facilities in the 1990s, during which I was working at the National Institute for Basic Biology as the director general. I was simply shocked to realize that the percentage of budget allocated to the biology field was really very small compared with those allocated to the fields of physics and chemistry. Furthermore, the budgets in other countries allocated for physics and other departments were comparable to those in Japan. In other words, the budget allocated to the field of biology in Japan was the smallest among developed countries. As a result, I took action to appeal to the then-Council for Science and Technology Policy, presenting these data so that they might consider improving the actual circumstances of the biased budget allocation. Since then, a significant portion of the budget has been allocated to researches on brain science and, with the disciplined reorganization of the engineering and pharmaceutical sciences into the life sciences, the low percentage allocated to biology seems to have been substantially improved. The total amount of the budget allocated to all of the disciplines in the life science field is still less than that allocated to other fields. Focusing on all of the natural sciences, even smaller amounts tend to be allocated to basic researches than those allocated to applied researches. I fear that this may lead to a further reduction in some basic researches.

Returning to 1980, the former members, including myself, of the Misaki Marine Biological Station held a meeting in honor of Ichiro Tomiyama, the former director, one evening. At the meeting, we had a very interesting conversation: "Nobel Prizes related to the natural sciences - the Nobel Prize in Physics, the Nobel Prize in Chemistry and the Nobel Prize in Physiology or Medicine - have been established, but not one for Biology, which would be awarded to biologists, including taxonomists. This may have an influence on peoples' academic rating of biological researchers, even though the Emperor, the symbol of the unity of the people, is a world-famous biologist. It would be better to establish an "Imperial Prize," to be given to researchers on basic biology, who are highly esteemed all over the world." In short, it was suggested that the Nobel Prize in Biology should be established in Japan. It was seven years later that Tonegewa won the Nobel Prize in Physiology or Medicine.

Early the following year, Tomiyama said, "By all means, let us make our fervent wish, namely, the establishment of the prize, come true"; unfortunately, however, he passed away a short time later, and the initiative for this project appeared to fade away. A few years later, under the leadership of Itaru Watanabe, a molecular biologist, who was Tonegawa's former mentor, researchers in the biological field brought the initiative back to life. I also visited, along with Watanabe, the relevant authorities from the then Ministry of Education, Gaishi Hiraiwa, the former chairman of Keidanren (the Japan Business Federation), Yoshihiro Tokugawa, the deputy grand chamberlain, and others, to petition them for the establishment of the prize. Nobuo Egami and Eiji Kobayashi, both former presidents of the Zoological Society of Japan, Hiroshi Hara and Makoto Numata, both former presidents of the Botanical Society of Japan, and Kiyoshi Aoki, a professor and the director of the Life Science Institute at Sophia University, where Watanabe worked, greatly contributed to our activities, in the background.

In 1983, affected by the strong enthusiasm in our demand conveyed through the Chief Cabinet Secretary, Takao Fujinami, the Science and International Affairs Bureau of the Ministry of Education, finally started working out a tentative plan for an "International Royal Prize." The Japan Academy was appointed as the authority responsible for awarding the prize, and it was decided that 10 million yen would be given to the winner as the supplementary prize. Headed by the then President of the Japan Academy, Hiromi Arisawa, some of its members, Seiji Kaya, Hitoshi Kihara, Naohide Hiratsuka, and others, dedicated efforts toward the realization of this plan. In this way, the plan was approved at a cabinet meeting through a process of negotiation with the government agencies concerned, academic societies and Keidanren, ultimately bearing fruit, namely, the establishment of the "International Prize for Biology" with the aim of "commemorating the 60th anniversary of His Majesty Emperor Showa's accession to the throne and his long-standing researches on biology, as well as encouragement of the development of biology" on April 25, 1985. It is said that the term "Royal" was not used in the English Prize name, in accordance with the Emperor Showa's will. The secretariat for the International

Prize for Biology is the "Japan Society for the Promotion of Science" (JSPS), which was founded using an Imperial donation from Emperor Showa in 1932 with the aim of promoting the sciences and has provided services such as international exchange and the selection of candidates to whom Grant-in-Aid for Scientific Research funded by MEXT is to be given.

"The Committee on the International Prize for Biology," consisting of representatives from various disciplines in the scientific field, was formed, with the President of the Japan Academy appointed as the chairman, to oversee this prize. The articles of the Committee include a provision that the "Selection Committee" shall include several foreign members, who, needless to say, remain anonymous. Foreign members are seldom included in the selection committees of the Nobel Prize (excluding the Nobel Peace Prize) and major prizes in Japan. As one of the features of this Prize, the field of specialization varies from year to year, so that researchers in disciplines that are important but not as widely recognized may also have equal opportunity to win the Prize. The Prize is awarded to only one researcher. Some have suggested that the Prize could be awarded to more than one researcher on a case-by-case basis; however, thus far, the "one recipient" rule has been followed.

The very first Prize was awarded to a researcher in the taxonomy or phylogeny field, to which Emperor Showa had devoted himself, and ever since, the winner has been selected from this field on every 10th anniversary of the prize's founding. The fund for the Prize has been covered by soliciting a wide range of communities for contributions, and those interested, including Keidanren, have contributed 500 million yen to the fund and continue their funding (quoted from a book written by the author [43]). Yasuhiro Nakasone was the Prime Minister when the establishment of the International Prize for Biology, as well as that of the Japan Prize, was approved by the cabinet. The latter was established through the funding of Konosuke Matsushita, the founder of Panasonic, in response to a governmental suggestion that it would be better to establish an international prize in Japan comparable to the Nobel Prize, to be awarded to those who have greatly contributed to the development of science and technology, with a supplemental prize of 50 million yen, with the aim of regaining the favor of the international community.

The following story concerning Masashige Kusunoki, a royal vassal in the Period of the Northern and Southern Dynasties (1336–1392), has been passed down. He was called "Tamon-maru" ("erudite child") in his childhood. An intelligent child, he practiced Buddhist asceticism at the Tennoji temple, Osaka. One day, a group of trainee monks at Tennoji, testing his wisdom, pointed at a temple bell and said, "Move this with one finger." After standing there thoughtfully for a while, he started to press his finger against it, with his feet firmly planted on the ground. Not surprisingly, it would not move an inch. However, after he had repeated this process several times, the large bell started to move bit by bit, finally swaying from side to side. The people watching admired him for his wisdom. The reason why I tell this old story here is that I had the same kind of experience as that in this historical episode. The topic discussed at the private meeting progressed in a beneficial direction by virtue of the efforts made by those from a variety of fields who appreciated the idea. Great was my delight when, a few years later, the idea was finally realized.

The ceremony of the International Prize for Biology is held on the last Monday in November or at the beginning of December at the Japan Academy, Ueno. Both Emperor Akihito and Empress Michiko have attended the ceremony since his Crown Prince days. Public attendance by Their Majesties is, aside from the International Prize for Biology, limited to two ceremonies for, respectively, the Japan Academy Prize and the Japan Prize and one conference of an international academic society selected from year to year. The Emperor used to deliver his "Okotoba" (message) after the report on the process of selection, until 2008. As described earlier, the Emperor himself polished his "Okotoba," throughout which his thoughts as a scientist were sprinkled here and there. Congratulatory addresses are delivered by the Prime Minister and the Minister of MEXT. Yasuhiro Nakasone and Yukio Hatoyama both attended the ceremony to deliver their speeches as Prime Minister, but ever since, the speech has been read on behalf of the Prime Minister due to Diet duties. The Minister of MEXT has attended almost every annual ceremony at which the medal and supplemental prize are awarded to the winners. As described earlier, the design of the medal features a colony of hydrozoans, *Pseudoclathrozoon cryptolarioides*, discovered by Emperor Showa. Besides this, a chrysanthemum-crested silver vase, the gift bestowed by the Emperor, is given to the recipients (Fig. 4.1).

After the ceremony, a buffet-style party is held to celebrate the recipient's brilliant feat in the presence of Their Majesties the Emperor and Empress. The Emperor always looks forward to attending this party and often stayed there far longer than scheduled in his Crown Prince days. However, following his enthronement, he began to leave as scheduled, in consideration of the security personnel accompanying him.

I cannot help but bring to mind the impressive scene when, one year, Empress Michiko, suffering from a lost voice due to psychological stress caused by denigration of her by the media, nonetheless chose to attend the ceremony and the celebration party afterwards together with the Emperor, who conveyed her thought to the attendants on her behalf.

The winners had always been invited to the Imperial Household during the reign of Emperor Showa, a practice that continued for a while after the name of the era was changed to Heisei, but this practice was halted because the schedule could not be made compatible with the symposium held for lectures on the selected field of biology, including the winner's commemorative lecture.

Recipients of the International Prize for Biology

Who has won the Prize so far, and for what subjects? Let us take a brief look at the recipients through reference to the booklet published every year by the Nomination Committee of the International Prize for Biology, though it might take a while. By

Fig. 4.1 The medal and a
silver vase (Imperial gift)
bestowed upon the recipient
of the International Prize for
Biology (Courtesy of the
Japan Society for the
Promotion of Science)

the way, the positions and nationalities listed for the recipients are those that were in
effect as of their reception of the prize [6–8].

1st (1985) in "Taxonomy or Systematic Biology": Edred John Henry Corner
(Professor Emeritus, University of Cambridge, UK)

He studied plants and fungi in tropical regions, especially in Southeast Asia. The theory of the co-evolution of animals and plants that he proposed, called the Durian Theory, is recognized to be the basic theory that underlies modern systematic phylogeny. During World War II, he served as the assistant director of the Botanic Garden in Singapore (Shonan at that time). It is a proud historical fact that, during his service, a joint effort by Japanese and English scientists, a group that included him, managed to preserve precious cultural properties with the generous cooperation of the then Marquis Yoshichika Tokugawa, and their collaborations continued even under the tragic conditions of wartime. In the speech at the prize award ceremony, the Emperor (the then Crown Prince) mentioned the book "Memorable Marquis: A Tale of Shonan-to (Singapore)" written by Corner, in which he himself described this episode.

2nd (1986) in "Systematic Biology and Taxonomy": Peter Hamilton Raven (Director, Missouri Botanical Garden, and Engelmann Professor of Botany at Washington University, US)

He investigated the issue of the diversity of plants by introducing a wide variety of methods in biology. He also investigated the co-evolution of insects and flowers with scientifically sound approaches and established the biology of pollination as the basis of evolutionary biology. He served for many years as the director of the Missouri Botanical Garden, which has the longest history of its kind in the US and was developed into a research center for tropical plants. Later, he was also awarded the International Cosmos Prize.

3rd (1987) in "Developmental Biology": John Bertrand Gurdon (John Humphrey Plummer Professor, University of Cambridge, UK)

Using amphibians, he injected a nucleus into the cytoplasm of an egg and first demonstrated that even the nuclei of once fully differentiated cells can be initialized and can reacquire the capacity to form larvae and adults, through a repeated developmental program. In 2012, the Nobel Prize in Physiology or Medicine was awarded jointly to him and Shinya Yamanaka, who discovered iPS cells that acquired pluripotency through initialization with a limited number of genes.

4th (1988) in "Population Biology": Motoo Kimura (Professor Emeritus, National Institute of Genetics, Japan)

He significantly contributed to the modern development of the theories of population genetics through his work on the "Diffusion Model." He also proposed "The Neutral Theory of Molecular Evolution," in which he proposed that Darwin's theory of Natural Selection alone is not sufficient to explain the evolution of organisms and that the random fixation of mutations neutral to selection in a population is also important. This theory had a definitive impact on the field of population genetics.

5th (1989) in "Marine Biology": Eric James Denton (Former Director, Marine Biological Association Laboratory, Plymouth, UK)

He made a series of new findings on the light adaptation of fish and clarified the relationship between physicochemical properties of ocean or sea water and the structure and function of marine animals, mainly through physical methods. Those achievements provided fundamental evidence for an easy understanding of the ecology and behavior of marine animals.

6th (1990) in "Behavioral Biology": Masakazu Konishi (Bing Professor, California Institute of Technology, US)

He unveiled many new facts to support the neuronal mechanisms of birdsong development in reproductive behaviors, as well as sound localization in owls in hunting their prey, through the use of original methods for behavioral analyses, as well as neurophysiological and anatomical methods. These not only developed the field of neuroethology, but also opened up a new field called Information Theory in Neuroscience. Currently a US citizen, he graduated from Hokkaido University in Japan.

7th (1991) in "Functional Biology of Plants": Marshall Davidson Hatch (Chief Research Scientist, Division of Plant Industry, CSIRO, Australia)

Based on experiments using sugar cane, he elucidated that C4-dicaboxylic acids such as malic acid, an initial product of carbon dioxide fixation in photosynthesis, are metabolized by the Calvin cycle through carbon dioxide fixation by the newly discovered C4-dicarboxyl acid pathway. C4 plants that have this pathway are more efficient at photosynthesis than common C3 plants, and therefore attempts are being made to introduce C4 functions into C3 plants.

8th (1992) in "Comparative Physiology and Biochemistry": Knut Schmidt-Nielsen (James B. Duke Professor of Physiology, Duke University, US)

He elucidated physiological mechanisms of various animals that inhabit severe environments, such as the mechanisms of water regulation and body temperature regulation of mammals and birds that live in deserts and those of salt and water regulation of marine birds and reptiles. Based on these findings, he revealed that those animals have enhanced systems that are common beyond species that help them to adapt to the environment. He was born in Norway and educated in Denmark.

9th (1993) in "Ecology": Edward Osborne Wilson (Professor of Science and Curator of the Entomology Museum of Comparative Zoology, Harvard University, US)

He studied Formicidae (ants) from a broad array of aspects, including ecology, biogeography and behavioral science, and brought a wide range of new knowledge about their colony structures, distribution, caste differentiation, chemical communication, and so on. He was one of the earliest biologists to point out that biodiversity is the most important issue, not only academically, but also socially. He also proposed a theoretical model that explains the difference in species diversity among organisms on islands. Later, he was awarded the International Cosmos Prize.

10th (1994) in "Systematic Biology and Taxonomy": Ernst Mayr (Professor Emeritus, Harvard University, US)

He is, undisputedly, one of the leading taxonomists and evolutionary theorists of the 20th century. Through the classification of birds, in regards to the concept of species that is a fundamental issue in taxonomy, he introduced "The Biological Species Concept," which not only takes morphology into consideration, but behavioral properties as well. Furthermore, he went on to develop it into "The

Theory of Allopatric Speciation," in which he proposed that species differentiation is triggered by reproductive isolation as the result of an increase in genetic variation of species due to geographic isolation.

11th (1995) in "Cell Biology": Ian Read Gibbons (Professor, Kewalo Marine Laboratory, University of Hawaii, US)

He clarified the ultrastructure of the cilia and flagella that are, aside from muscle, responsible for motion within the animal body and identified a motor protein dynein that is an essential element of the structures. He also conducted a series of experiments to prove the mechanism of their motions and contributed to studies that led to the resolution of important issues related to cell motility and cytoskeletons using an electron microscopic and both biochemical and molecular biological methods.

12th (1996) in "Biology of Reproduction": Ryuzo Yanagimachi (Professor, Medical School, University of Hawaii, Japan)

He conducted a number of pioneering researches in the field of reproductive biology, especially on fertilization in mammals, and made significant contributions to advances in two fields, basic science and applied science. One example of his contributions is the success of in vitro sperm activation (called capacitation). Without benefit of the fruits of this research, Robert Edwards, who is well known for creating the first test-tube baby, would not have been able to win the Nobel Prize in Physiology or Medicine in 2010. Yanagimachi also blazed a path to the present fertility treatment technology that enables sperm nuclei to be injected into the cytoplasm of an egg (Intracytoplasmic Sperm Injection, ICSI) so as to induce embryonic development. He graduated from Hokkaido University, Japan, and maintained his Japanese nationality at that time.

13th (1997) in "Plant Science": Elliot Martin Meyerowitz (Professor, California Institute of Technology, US)

Using samples small in genome size of *Arabidopsis thaliana*, which belongs to the Brassicaceae family, as a model plant, he established approaches to elucidation at the molecular level of the development and differentiation of plants, especially tissue and organ formation, and of the biological phenomena specific to plants, such as their unique metabolism, achieving rapid progress in plant molecular genetics. He is particularly well known for the ABC Model, which explains the mechanism of flower development by the combinatory action of three regulatory genes.

14th (1998) in "The Biology of Biodiversity": Otto Thomas Solbrig (Bussey Professor of Biology, Harvard University, US)

After dedicating himself to the field of plant diversity and evolution, racking up outstanding achievements over the years, he urged the importance of integrated biodiversity research on a global scale and played a leading role in promoting international collaborations between the International Union of Biological Sciences (IUBS) and UNESCO.

He graduated from the University of La Plata, Argentina.

15th (1999) in "Animal Physiology": Setsuro Ebashi (Chair of the Section II, The Japan Academy, and Professor Emeritus, University of Tokyo, Japan)

He contributed significantly to the elucidation of the mechanism of skeletal muscle excitation and contraction triggered by the stimuli from motor neurons, which had been a big question in animal physiology, through the discovery of the uptake of calcium ions by the sarcoplasmic reticulum and involvement of the troponin-tropomyosin system, as well as identification of actin-binding proteins. He was co-nominated many times for the Nobel Prize in Physiology or Medicine with Hugh Huxley, who proposed the "Sliding Filament Theory" of muscle contraction.

16th (2000) in "Developmental Biology": Seymour Benzer (James Griffin Boswell Professor of Neuroscience, California Institute of Technology, US)

He treated fruit flies, *Drosophila melanogaster*, with chemicals and artificially produced mutants that displayed various forms of abnormal behavior, investigating defects in their sensory organs, central nervous systems, and motor systems through genetic dissection. Through these studies, he made pioneering contributions to the elucidation of the genetic mechanism of the nervous system and the mechanisms of neural development and differentiation.

17th (2001) in "Paleontology": Harry Blackmore Whittington (Professor Emeritus, University of Cambridge, UK)

He studied the body plan, ecology and evolution of trilobites for many years, advancing biological knowledge of these bygone creatures to remarkable degrees. He also carried out a reexamination of the unique Burgess fauna of the middle Cambrian and shed light on the nature of the giant predator *Anomalocaris*; these revealed the whole picture of the "Great Cambrian Explosion," an extremely important event in the history of terrestrial organisms. He majored in geological science.

18th (2002) in "Biology of Evolution": Masatoshi Nei (Evan Pugh Professor of Biology, Pennsylvania State University, US)

He played a central role in establishing the theoretical basis for the present form of molecular evolutionary biology. In order to study the genetic diversity of a group of living organisms and the evolutionary relationships among different species of living organism at the molecular level, he invented novel statistical methods, such as methods for accurately estimating the time when a species branched and for searching for genetic regions that are subject to natural selection. The "Neighbor Joining Method" of phylogenetic tree construction that he invented is especially frequently used worldwide. He is a graduate of the University of Miyazaki. Later, he was awarded the Kyoto Prize.

19th (2003) in "Cell Biology": Shinya Inoué (Distinguished Scientist, Marine Biological Laboratory, Woods Hole, US)

He invented a "Polarizing microscope" of his own original design (often called a Shinya-scope, with respect to the prize recipient), and observed spindle fibers (microtubules) during the cell division of live cells for the first time. Furthermore, he greatly contributed to a recent advance in the dynamic imaging of live cells and cell components by enabling us to see a single macromolecule under the imaging system using a video camera, etc. He is a graduate of the University of Tokyo.

20th (2004) in "Systematic Biology and Taxonomy": Thomas Cavalier-Smith (Professor, Department of Zoology, University of Oxford, UK)

Based on his specialized knowledge of cell biology, electron microscopy and molecular biology, as well as on the latest knowledge of all areas of biology, including physiology, biochemistry, and so on, he precisely, yet boldly, organized and systematized the higher order phylogeny of living things. It was he who advanced a new theory that fungi are closely related to animals. He added one new kingdom to the preexisting five kingdoms theory, proposing a six kingdoms theory instead, and attempted to establish a more natural taxonomy with no intentionality.

21st (2005) in "Structural Biology in Fine Structure, Morphology and Morphogenesis": Nam-Hai Chua (Professor, Laboratory of Plant Molecular Biology, Rockefeller University, US)

He discovered the "Transit Sequence" required for the transportation of proteins to chloroplast, identified the organelle for photosynthesis in plants, and elucidated the gene induction mechanism of such proteins by lights. Furthermore, he conducted advanced studies related to plant morphogenesis and revealed the involvement of Auxin in the body axis formation of plants and the molecular mechanism of lateral root formation. He graduated from the National University of Singapore.

22nd (2006) in "Chronobiology": Serge Daan (Professor, Niko Tinbergen Chair in Behavioral Biology, University of Groningen, the Netherlands)

He investigated the mechanisms of behavior and sleep of various animals, including humans, not only through experiments in the laboratory, but also through examinations and experiments in the field. With these studies, he elucidated the fundamental role and mechanisms of two biological rhythms, circadian rhythm (a cycle of approximately 24 h) and yearly rhythm (a cycle of approximately 365 or 366 days), and established the basis for chronobiology. The outcomes of his studies have been applied to treatments of seasonal depression and jet lag (Fig. 4.2).

Fig. 4.2 Their Majesties attending the International Prize for Biology award ceremony in 2006 (Courtesy of the Japan Society for the Promotion of Science)

23rd (2007) in "Genetics": David Swenson Hogness (Munzer Professor of Developmental Biology and Biochemistry, Emeritus, Stanford University School of Medicine, US)

He first succeeded in making a DNA library of *Drosophila*, one of the higher eukaryotes, and demonstrated the presence of the TATA box that controls gene expression. He promoted research on the structure and function of genes and their regulatory mechanisms and racked up many achievements that have provided the basis for our current understanding of genes. He also presented the first experimental evidence to show that gene functions are indeed necessary for animal morphogenesis.

24th (2008) in "Ecology": George David Tilman (Regents' Professor, Department of Ecology, Evolution, and Behavior, University of Minnesota, US)

He made theoretical predictions about the way in which biodiversity is established and maintained and how it contributes to the function and stability of ecosystems, as well as carrying out long-standing experiments in a substantial number of experimental grass fields with varied species diversity for more than 10 years. He demonstrated that, with higher levels of diversity, the stability of an ecosystem is more greatly enhanced. He has also promoted researches related to sustainable agriculture and biofuel.

25th (2009) in "Biology of Sensing": Winslow Russell Briggs (Professor Emetitus, Department of Plant Biology, Carnegie Institution of Washington, US)

He has studied the mechanism (photoreaction) by which plants react to light information for many years and, in relation to phototropism, he discovered a blue light receptor protein, phototropin, which is the receptor required for plants to recognize the direction of light. With this achievement, he significantly contributed to researches on the photoreaction of a wide variety of organisms ranging from bacteria to seed plants.

26th (2010) in "Biology of Symbiosis": Nancy Ann Moran (William H. Fleming Professor, Department of Ecology and Evolutionary Biology, Yale University, US)

She studied the close relationship between insects and the symbiotic bacteria that reside within their bodies, making full use of a wide variety of approaches, including molecular biology, genome science, experimental biology and theoretical biology. She also shed new light on issues such as the evolutionary origin of the symbiotic relationship, universality in the animal kingdom, host-symbiont co-evolution and the possible impact of symbiosis on ecology. She was the first female recipient of the prize.

27th (2011) in "Developmental Biology": Eric Harris Davidson (Norman Chandler Professor of Cell Biology, California Institute of Technology, US)

Consistently using the sea urchin as an experimental model, he addressed central issues of developmental biology, such as the function of genes in development, the regulatory mechanism of gene expression, the cascade of gene expression and the mechanism of development and evolution. He amalgamated the findings of his studies through the concept of a "Gene Regulatory Network," and also succeeded in its experimental demonstration.

28th (2012) in "Neurobiology": Joseph Altman (Professor Emeritus, Purdue University, US)

Although cell division and differentiation continue over the lifetime in most of the body's tissues, it has been believed for many years that this is not true of the brain tissues, and that the neuronal circuit, once injured, can never be regenerated. However, Altman experimentally demonstrated, for the first time, that new neurons could be born in the brain of an adult mouse. This fact has recently been successfully verified and has opened up the possibility of treatments for regenerating the nervous system.

29th (2013) in "Biology of Evolution": Joseph Felsenstein (Professor, University of Washington, US)

Based on the data of allelic genes, etc., he invented the "Maximum Likelihood Method" for constructing phylogenetic trees of populations. He further developed a new algorithm of a method for constructing phylogenetic trees of genes based on their DNA sequences, packaged the program as the "PHYLIP" software, and disclosed it to the community. He also introduced the bootstrap value to confer reliability on the branching of phylogenetic trees. Through these achievements, he has significantly contributed to the field of evolutionary biology.

30th (2014) in "Systematic Biology and Taxonomy": Peter Crane (Professor, Yale University, US)

The origin and early evolution of land plants, especially of angiosperms, have long been unsolved issues in the field of biology. Crane first succeeded in investigating them by integrating paleontological studies based on fossil plants and comparative morphological studies of extant plants, thereby opening up a new avenue for plant taxonomy.

31st (2015) in "Cell Biology": Yoshinori Osumi (Honorary professor, Tokyo Institute of Technology, Japan)

He cast light on the phenomenon of "autophagy," which had only previously been recognizable under an electron-microscope, and pioneered studies on the molecular mechanism of the process for protein recycling, taking advantage of the genetics of yeast. He and his colleagues also revealed that the system is highly conserved among species, including humans, which implied its relationship to the development, nutrient starvation, and pathogenesis of a variety of diseases. He was awarded the Nobel Prize for Physiology or Medicine in 2016 for this outstanding achievement.

32nd (2016) in "Biology of Biodiversity": Stephen Philip Hubbell (Distinguished Professor, University of California, Los Angeles, US)

He greatly contributed to biodiversity research by proposing the unified neutral theory of biodiversity and biogeography, which he then evaluated through large permanent forest census plots of tropical and temperate forests around the world. His achievements, ranging from theoretical works to global-scale field works on the diversity of forest trees, have had a strong impact on biology and are widely appreciated.

33th (2017) in "Marine Biology": Rita Rossi Colwell (Distinguished University Professor, University of Maryland, College Park, and John Hopkins Bloomberg School of Public Health, US)

She contributed significantly to the refinement of the taxonomy of marine bacteria by introducing genetic analyses and bioinformatics using advanced computing technologies, which established the taxonomy of the genus Vibrio, including *V. cholerae*, which causes cholera. She also studied the ecology of vibrios and uncovered unique habitats of theirs that depend on water temperature, which suggested a close relationship between global warming and the range expansion of vibrios.

These names may not be familiar to those outside of the scientific community, but all recipients were carefully nominated after active discussions among national, as well as international, authorities, including the Nobel laureates, and they surely deserved the Prize, truly being the top scientists in their individual fields. The contents of the researches for which each prize was given do indeed reflect the recent advances in modern biology.

Hoping to Shed More Light on Science, Biology and Basic Research

The result of the 2006 Survey for International Student Assessment (PISA) showed that the 15-year students in Japan did not think that science would give them a good chance for a successful life, unlike those in other countries, and the rate of students who predicted that they would be working in any type of scientific job at the age of 30 was only 8%, compared with 25%, the average of the OECD member countries, putting Japan at the lowest rank. Although they get good scores on the PISA test, the students feel the lowest level of confidence in their scientific ability among OECD member countries. On the other hand, the percentage of high school students who think that science will mostly be useless to them when they going out into the world is the highest. Why is that? It is an incomprehensible phenomenon in view of the recent situation in which female scientists are lionized and the promotion of science and technology has been called for.

Moreover, in 2002, the National Institute of Science and Technology Policy carried out a survey with a questionnaire on basic scientific knowledge among 2,000 persons between 18 and 68 years of age. A comparison with the percentages of correct answers on the questionnaire among 14 advanced countries showed that Japan ranked 12th. However, according to the Trends in International Mathematics and Science Study (TIMSS) carried out in 2000, the average result of junior high and high school students in Japan was ranked 2nd among 31 countries, suggesting that a large gap exists between the above two results. This means that, since the youth study only to ensure their success on entrance examinations in their junior high and high school days, they will completely forget any knowledge learned when they grow up, losing interest in science. In contrast, in the US, the scores of students in junior high and high schools are poor in science, while the adults have acquired scientific knowledge at the highest level.

The unfortunate results of the surveys may be partially attributable to the Japanese character, but it is believed that there are also problems with the university entrance examination and education systems, as well as with the attitudes of public authorities and industries. Most people who strive to enter the political and business worlds and the mass communication media have a background in the humanities. For this reason, the students who intend to study subjects in the humanities tend to begin to become disassociated from science as early as junior high school, and even more become so in high school, because questions related to these subjects are not featured in the examinations for university entrance. Furthermore, it must be acknowledged that, recently, the Department of General Education, which had been established to facilitate students in the country-wide universities acquiring broad knowledge in the liberal arts and sciences, was broken up, aggravating that tendency, although some would say, "We have to follow the policy." The lackadaisical attitude of the Prime Minister of Japan, who says "I have no knowledge of science at all…," is questionable. Favorably, it has been suggested that he might at least keep advisors steeped in the sciences around him, in addition to those steeped in politics and economics. It is obvious that people with scientific backgrounds need to do a better job mastering knowledge of the broader culture. From this standpoint, Emperor Akihito and Prince Akishino, both of whom had studied subjects in the humanities up to university, despite their early interest in animals, attained excellent achievements as biologists and a deep all-around knowledge of science. We should learn from their attitudes.

At the present time, the term "scientific technology" is commonly used in Japan. This tends to refer to any technology, technique or industrial mechanism. On the other hand, it is also translated into "science and technology" in English. A serious problem lies hidden within this situation. Ideally, it should be translated into "science and technology" instead of "scientific technology," or at least "science/technology," where appropriate. However, the mass media and the Japanese public have never addressed this problem. Although the former Ministry of Education, Science and Culture (before its merger with the Science and Technology Agency) introduced the word "academic research" in place of "science," it seems that the public's acceptance of this representation has not gone smoothly. In Japan, since around the time of the collapse of the bubble economy, the need to encourage practical and applied researches has been advocated with the slogan of a "Science-and-technology-oriented nation." The government, the Diet members and the mass media have united in their common interest, shouting about the need for researches to be immediately applied to this problem or that issue. This implies that top priority must be given to cost-effectiveness; that since science is financed by taxes, "useless" researches should be excluded. The result of this situation is that fundamental researches intended to seek the truth have been forced to scale down, or, at best, maintain the status quo, with the exception of applied researches. The conversion, since implemented, of national universities and research institutes into corporations has reduced subsidies by 1% every year, increasingly aggravating this tendency.

How can Japan, whose ratio of budget allocated to institutions of higher education to GDP is the lowest among OECD member countries, achieve a "science-and- technology-oriented nation"? A source from the Ministry of Finance apparently considers that there is no problem, because, in their estimation, a perfectly adequate amount of the budget has been allocated already; however, it's a depressing prospect in terms of the future. It is definitely true that the competitive research funds have been increased, but the budget allocated to fundamental research programs with a strong potential for dramatic development in the future, allowing researchers to attempt to explore new territories with no restriction, has been considerably reduced. No matter what efforts are made to conduct advanced researches, the attitude in regard to following the example of foreign countries would only cause existing researches and deliverables to be improved, and would not bring about great progress. As thoughtful researchers, including the Nobel Prize winners, have repeatedly insisted, cases in which research results that had initially seemed to have no immediate application eventually led to a great breakthrough after years or even decades are too numerous to mention. But before such research results can be achieved, the broad bases for their related fields need to have been established. It stands in sharp contrast that the competitive research funds have increased, whereas the research funds that a single researcher can use at his or her discretion were reduced to 100,000–200,000 yen from 500,000 to 1,000,000 yen. This means that, with no Grants-in-aid for Scientific Research given, researchers could not do their work. Even in cases in which they failed to get their grants, the possibility of them at least continuing to work on their research, which may serve as a seed for successive researches, must be taken into consideration.

As described earlier, as of 2017, the number of Japanese winners of the Nobel Prize in the fields of the natural sciences is 11 in Physics, 7 in Chemistry and 4 (plus 1 in 2018) in Physiology or Medicine, including those who are now US citizens. The number of winners in the top-three countries, the US, the UK and Germany, are 92, 25 and 24 in Physics; 68, 28 and 29 in Chemistry; and 102, 30 and 16 in Physiology or Medicine, respectively. Compared with those in these three countries, the number of Japanese winners in Physiology or Medicine is small (only 2 when the original edition of this book was written). "Ordinary researchers," so to speak, in Japan have been able to win the Nobel Prize in Physics or Chemistry at a level similar to that in the US, but the Prize in Physiology or Medicine has stayed beyond Japanese researchers' reach. As described earlier, this is the result of various kinds of unfortunate circumstance. The good news, however, is that more than one researcher has been nominated for the Prize in Physiology or Medicine in recent years. We biologists need to make serious efforts toward obtaining that Prize.

On the other hand, it is a clear and indisputable fact that a reduced amount of funds has been allocated to researches on biology, as described earlier. Considering that many patents for major medicines have been granted to foreign companies, and that both the US and the UK have begun putting major focus on the life sciences, the field would seem to be an obvious one for Japan to allocate a greater portion of the budget to in the future.

By the way, extra-large facilities, such as the high intensity proton accelerator built in Tokai-mura, Ibaraki, at a cost of 150 billion yen, and the Extra Large Telescope being built on Mauna Kea, Hawaii, at a cost of about 40 billion yen, are not always necessary for researches on biology (note that the above-mentioned facilities are also used for researches related to the life sciences). The Human Genome Project, which was initiated by Japan but ended up upsetting many molecular biologists with its late start, required a large number of essential sequencers, but only cost about 10 million yen per unit. Needless to say, there are various types of equipment that cost several hundred million yen. However, generally, the cost per unit is lower than that of units for physical or chemical experiments. Experiments on organisms are most costly in regard to the researchers' expenditure of labor and time. For example, iPS cells may not yield practical applications before the researchers have carried out their experiments on mice repeatedly over a long period of time, or, in some cases, over several generations, to demonstrate that they would not produce cancers. Naturally, this process requires someone to raise mice for experimentation during the preparation and experimental periods. To address challenges such as environmental conservation and biodiversity conservation, even longer-term and steadier investigations and observations are required. In any case, the targets for experiment are living organisms. In recent years, as in the case of the fields of physics and chemistry, funds as high as several hundred million yen have been allocated to certain fields of biology on a preferential basis, however, I believe that better results would be achieved if roughly ten million yen in research funds were granted to the individual researchers.

Focusing on natural history, one of the academic areas in the field of biology in which the Imperial Family members have been earnestly engaged, the idea that a natural history museum would be capable of playing a central role in the promotion of the natural sciences in Japan has never taken root, and most taxonomy and phylogeny courses have been discontinued in Japan, as described earlier. Worse still, the abolishment of the Department of General Education deprived those taxonomists who still remained in the Department of a chance to develop their successors. On the other hand, the Ministry of Education, Culture, Sports, Science and Technology (MEXT) tried to shore up the "University Museum" as part of a policy of creating a natural history museum, resulting in the concentration of valuable materials, which had previously been dispersed around a university, into one archive. Moreover, as seen in the collaborative efforts between the National Museum of Nature and Science and the University of Tokyo, the postgraduate course for developing specialists in this area continues to survive in a small way. The taxonomists who developed new methods of molecular biology are now able to discuss the theories of phylogeny and evolution with firmer assurance than before, leading to a well-organized research environment for young researchers.

Nevertheless, looking at the entire field of biology and the life sciences, we would have to say that the area of taxonomy and phylogeny is one of conservative, fundamental researches. The researchers specializing in the subjects of taxonomy and phylogeny are themselves becoming "endangered species." If we are to acquire the basic data necessary to conserve biodiversity, which is essential to the survival of mankind, constant support for the natural sciences is absolutely imperative. In addition to the natural sciences, the individual fundamental areas of biology, in which we learn much from living organisms, may undoubtedly reap various types of benefit for the future of humankind, because we humans are also living organisms.

Finally, the Emperor Akihito, who pays attention to both the merits and draw-backs brought on by science, delivered the message at the opening ceremony of the International Nuclear Physics Conference held in 2007:

> Delineating an outlook for the 21st century, it may be necessary to pay attention to the historical process in which an advance in science decided the outcome.

Moreover, he spoke of the brilliant advances in physics in the 20th century, as well as the damage caused by weapons of mass destruction. In expressing the above-mentioned thought, he seems to be referring to all areas of science, not just those related to the atomic nucleus. In fact, problems of contamination are just as easily caused by insecticides and herbicides, even in the field of biology, and advances in reproductive technology and organ transplantation involve ethical problems. More fundamentally, the prosperity of mankind is itself causing mass extinctions of life on earth, exceeding the scale of that of the dinosaurs. Makoto Watanabe describes in his book:

> The underlying philosophy of the Emperor seems to be that the ultimate objective of science has to be ... to bring happiness to the people through "love for the people" [68]

It is possible that the Emperor's thoughts envelop all animals and living creatures.

Afterword

It is auspicious that both of Their Imperial Majesties, the Emperor and the Empress, turned 80 in 2014, 55 years after their marriage. In the same year, the International Prize for Biology, which was established in commemoration of the contributions to biology made by the Emperor Showa and the present Emperor, celebrated the 30th anniversary of its establishment. Furthermore, being worried about his possible failure to perform his official duties satisfactorily as a symbol of the unity of the people due to his advanced age, the present Emperor decided that he would abdicate the throne in April 2019. I began writing this book in hope that a great number of people would come to recognize the enthusiasm for and contributions to biology by the members of the Imperial family. In working on it, however, I was shocked to find that I myself, as a biologist, knew very little of the everyday concerns related to the Imperial Family members, including their individual contributions to science, before the commencement of this work, partially because of the fact that I was engaged in a different area in the field of biology. Having only poor knowledge of the Imperial Family members, I could not make the readers understand the naked truth about them. Therefore, I read as many of their original papers and books as possible, including relevant literature, and had many conversations with other interested parties.

Through this process, I gained solid recognition of how seriously they have devoted themselves to researches on the subjects in their own individual fields, taking time out from their extremely busy schedules. Although I have recently become involved in the area of taxonomy and phylogeny, as a non-professional in the area, I had to absorb a lot of knowledge in the process. Undoubtedly, the Imperial Family members have enjoyed the benefit of having the opportunity to become acquainted with the most eminent scientists and researchers of our time and elicit their support, while, contrastingly, they are subject to severe restrictions in finding the necessary time and space for their researches. Under such circumstances, before they were able to attain the achievements that were so highly

© Springer Nature Singapore Pte Ltd. 2019
H. Mohri, *Imperial Biologists*, Springer Biographies,
https://doi.org/10.1007/978-981-13-6756-4

esteemed by researchers in foreign countries, they must have had to put in an extraordinary amount of effort. Moreover, they are all so modest that none of them would think of making a big deal about their researches to others. Furthermore, the people around them also do not talk much about their researches, taking their desires into consideration. Accordingly, with the exception of Emperor Showa, it was sometimes difficult for me to collect relevant materials, as I usually had no opportunity to interact with the Family members, unlike their chamberlains and the reporters assigned to the Imperial Household.

Despite a number of them having been devoted to researches in the same area, taxonomy and phylogeny, one can observe unique characteristics among the processes that the individual members follow or have followed, reflecting the social conditions of the assorted eras. Emperor Showa, who was brought up to be the "Arahitogami (living god)," intended to stand firm in his position as a constitutional monarch and was referred to as the last Emperor of Japan, had his own collecting boat and used a warship in some cases, despite his usual modesty. As is known from his descriptions of new species, he collected a wide range of plants and animals within limited areas in Japan, specifically Sagami Bay, Nasu and Suzaki, and he specialized in hydrozoans and myxomycetes. He put the publication of new species that he discovered in the hands of other cooperative specialists, excluding those of hydrozoans and the plants that he collected in Nasu, Suzaki and the Imperial Garden, which were published under the co-author's name, Hirohito. Most of them were published as special editions released by the Biological Laboratory of the Imperial Household.

Emperor Akihito reached the age of discretion under the Japanese constitution established in 1946, which states that the Emperor shall be recognized as the symbol of the unity of the people, and has been unwilling to accept preferential treatment over the citizens of the nation. As a result, he has carried out his study of fish energetically and with deep seriousness, especially in regard to the family Gobidae, and has written and contributed papers to academic journals, in which submitted papers are peer-reviewed, all by himself. He has not only dissected fish, but has also drawn his own illustrations for his papers and encyclopedias of fish. He assumed new morphological features of fish to be indicators in taxonomy and phylogeny, but has recently initiated DNA-based phylogenetical analyses and is trying to spur discussions of the problems from both the morphological and molecular phylogenetical viewpoints.

There are differences among the research support systems surrounding Emperor Showa, Emperor Akihito and Prince Akishino. Reflecting the social conditions of the time, a number of scientists, most of whom graduated from the Department of Zoology or Botany, University of Tokyo, made great contributions to Emperor Showa's work. However, as time went by, the number of taxonomists, especially those working on marine organisms, was reduced at the University of Tokyo. For this reason, specialists working separately at local universities or institutes provided support for his taxonomical research on the specimens of animals and plants collected in Sagami Bay. The "Observers of the Imperial Court School," including the grand chamberlain, also cooperated with him in collecting animal and plant samples.

Emperor Akihito was initially supported by Ichiro Tomiyama and Tokiharu Abe, both graduates of the Department of Zoology, University of Tokyo, with many other scientists engaged in fishery science, mainly from a network of personal connections from the Tokyo University of Fisheries (formerly the Imperial Fishery Institute and the present Tokyo University of Marine Science and Technology) making great contributions to his research. In his recent study of the organisms living in the garden of the Imperial Household, many scientists from the National Museum of Nature and Science and others joined in cooperation with him. Prince Akishino, on the other hand, has mainly carried out his joint researches with the scientists who work at institutes such as the Research Institute of Evolutionary Biology, SOKENDAI and the Yamashina Institute for Ornithology, as well as foreign researchers, including those in Thailand. He is also engaged in an active exchange with researchers majoring in the humanities, including those working at the National Museum of Ethnology. It is common to all of the Imperial Family members that they maintain an affectionate eye towards living organisms, as can also be seen in Prince Hitachi's cancer research and Sayako Kuroda's ecological studies on kingfishers. This is unquestionably because of their feeling of deep reverence for life. This genuine feeling is often expressed as deep sympathy for the war dead of World War II and for disaster victims, as well as consideration paid to conservation of the natural environment, including affection for animals designated as endangered species and plants living along the roadside.

Even though studying a different subject than biology, Crown Prince Naruhito, the successor to the throne, has been carrying out research on the way in which water and transportation are essential to everyday human life in the same steady and academic manner as the other family members have done. It seems that his background growing up in an environment surrounded by Imperial scientists has had a considerable effect on him. The academic community admires the Imperial Family for their wide range of knowledge, but their recognition is not necessarily correct. Although the Imperials understandably give top priority to their tight schedule of official duties and religious services, each of them is a specialist in their own academic area, though they are never willing to acknowledge that fact. In addition, they started by learning the basics of biology. Having been raised by his parents, both possessors of doctorates, the Prince Hisahito, who is expected to succeed to the Imperial throne at some future date under the current Imperial system, is fond of animals and may intend to become a fourth-generation biologist. I am greatly looking forward for that day to come!

At any rate, there is no such case in other countries in which an Imperial Family has engaged in academic disciplines, especially biology, from generation to generation and attained excellent achievements. Again, we should be proud of these achievements as being equal to or even at a higher level than those of Drs. Tonegawa, Yamanaka, Omura and Osumi (and Honjo), the Nobel Prize laureates in Physiology or Medicine. At the same time, we hope that more attention will be focused on the life sciences and fundamental sciences (including all academic disciplines) in which the truth is sought, in addition to the material sciences and applied researches, which have direct applications, and that great financial support will be provided for them. The very real fact that a huge amount of money has been invested in the life sciences in the US and the UK needs to be recognized. Moreover, the tradition of natural history has been carefully passed down in these countries. Three Japanese researchers who contributed to the development of LED, from which people all over the world enjoy benefits, won the Noble Prize in Physics in 2014. This is an admirable achievement! On the other hand, as mentioned previously, it has been said by successive Nobel Prize winners, one after another, that breakthroughs can be achieved from fundamental researches aimed at no direct applications.

More than 30 years have passed since the establishment of the International Prize for Biology. Taking advantage of its 25th anniversary, it was decided that further development of biology would be promoted in commemoration of the long-standing research on gobies by Emperor Akihito, who has made such great contributions to the development of the prize. Above all, this prize, rumored to be called the "Emperor Prize" in secret, enjoys moderate awareness among the general public, but is, unfortunately, not so widely known throughout the world. In fact, the mass media in Japan has given almost no coverage to this prize since its establishment.

TV and print journalists tend to show up for the award ceremony, but news of the prize tends to center around, for example, programs and articles such as "Koshitsu Arubamu" ("Album of Imperial Family"), showcasing the appearance by the Imperials, with the winner and the details of his or her work rarely ever being reported. The researchers who win the prize will occasionally be mentioned in some article unrelated to the prize; even in cases such as this, it is seldom if ever reported that they are winners of the International Prize for Biology. I wonder if such a situation can be allowed to continue. The mass media tends to flock to Stockholm with a great deal of fuss whenever Japanese researchers win the Nobel Prize and a great deal of space is given to any report about a foreign prize having been won. This is an undeniable tendency, and I cannot see how Japan expects to be taken as a cultured nation when no reports on news of this time-honored Japanese prize are ever written. The Nobel Prize has earned authority and renown, as we know, through the long-standing, steady efforts of the Swedes. I once wrote an article for a newspaper on this very subject, but it was not printed for the reason that so many other prizes with a roughly ten million yen award have been established.

In fact, the Inamori Foundation was established through a financial contribution by Kazuo Inamori of the Kyocera, in the same year as the International Prize for Biology, for the purpose of setting up the "Kyoto Prize" in order to honor those who have greatly contributed to the development of science and culture and the deepening and uplifting of the human spirit. This Prize focuses on three disciplines, among which the fundamental sciences include Bioscience and Life Science as the targets, with 50 million yen in prize money awarded per winner. Moreover, the "International Cosmos Prize", which is awarded to those "who are recognized for their great contributions to the creation and development of a philosophy of "co-existence between people and nature,'"' was established in 1993 with a 40 million yen award in commemoration of the "International Garden and Greenery Exposition (Flora Expo)" held in Osaka in 1990. Many prizes with a more than 10 million yen award were established. On the other hand, the prize money for the Fields Medal in mathematics, ranked with the Nobel Prize, ranges from only 1 to 2 million yen.

I hope that the prizes established in Japan will be further spread throughout the world, with the abandonment of such traditional attitudes as admiration for foreigners and pointless restraint. Naturally, we, the people concerned, have to put in the efforts to make it happen. Makoto Watanabe, the former grand chamberlain, has stated:

> Incidentally, His Majesty the Emperor often says, "the Japanese people tend to pay great attention to foreign prizes, including the Nobel Prize, but not to the prizes awarded in Japan, including the Japan Academy Prize. They should be regarded as more important."
> [68]

In a situation in which so many foundations have encountered difficulties in management due to the stingy monetary policy that has been enforced in recent years, the International Prize for Biology has similarly suffered from financial stringency. Owing to contributions from private enterprises, academic societies and private individuals, including recipients of the Prize, it has managed to deal with its immediate funding needs, although it has already been forced to use up those contributions. I fervently hope that those who wish to sustain or support and develop the Prize established in Japan to praise biologists throughout the world in commemoration of the two generations of Emperors, the symbols of the unity of people, who have themselves been serious biologists, will make a contribution, including through inheritance, to the Prize. To make a contribution, feel free to contact the Japan Society for the Promotion of Science (Koji-machi 5-3-1, Chiyoda-ku, Tokyo 102-0083, Japan). Following the proverb "Start from small things," I myself make a small contribution to the Prize every year.

It should be noted that, in 2003, the Norwegian Government established the "Abel Prize," with 100 million yen in prize money, named after a Norwegian mathematician, as an international prize to be awarded to mathematicians as an alternative to the Nobel Prize. No Japanese researcher has yet won this prize. In the UK, the Queen Elizabeth Prize for Engineering (QEPrize), with 130 million yen in award money, was established in 2013, and multiple Japanese enterprises have made significant contributions to the QEPrize foundation. This could definitely lead to effective promotion of such prizes, but I still have my doubts. We who are concerned with the future of the International Prize for Biology also have to make efforts to solicit contributions from foreign entities.

I am grateful to a lot of people for their help in writing this book. In particular, I would like to express my profound gratitude to Drs. Mayumi Yamada, Hiroshi Namikawa and Patricia Morse for their cooperation in writing the section concerning the Emperor Showa; to Makoto Watanabe, the former Grand Chamberlain, Katsusuke Meguro, the former Chamberlain, the staff of the Biological Laboratory of the Imperial Household, and Drs. Ryoichi Arai and Tetsuji Nakabo for their cooperation in writing the section concerning the Emperor Akihito; to Dr. Tomoyuki Kitagawa for cooperation in writing the section concerning the Prince Hitachi; and to Drs. Yasuhiko Taki, Tomoya Akimichi and Yoshihiro Hayashi for their cooperation in writing the section concerning the Prince Akishino. Moreover, I greatly appreciate the gracious suggestions given to me unexpectedly by the Empress at one stage of writing. I express my deep thanks to many persons for their kind consideration in providing invaluable details. I felt especially taken care of by Hironobu Sanmiya, the former editor-in-chief, and Yumiko Nara of the Asahi Shimbun Publications in publishing this book. Finally, I express my gratitude to my wife Toshiko for her long-standing support.

References[1]

1. Akashi M (2013) Dignity and sincere personality of the Present Emperor—Emperor Akihito told by his fellow student (Kinjo Tenno Tsukurazaru Songen – Kyuyu ga Tsuzuru Akihito-Shinno). Kodansha
2. Aoki J (2013) Course hour of natural history—Learning science from mother nature. University of Tokyo Press
3. Arai R (1990) His Majesty Emperor Akihito and his research on gobies—Imperial Family in the Heisei Period. In: A book of photographs commemorating the accession to the throne. Kyodo News Service
4. Asahi Shimbun Inc. (ed) (2009) Emperor Akihito tells natural science (Tenno-Heika Kagaku wo Kataru). Asahi Shimbun Publications Inc.
5. *Bayer FM (1988) Message. In: The National Museum of Nature and Science (ed) The Emperor's Biological research. The National Museum of Nature and Science
6. Committee on the International Prize for Biology (1994) The Ten Years of International Prize for Biology. Japan Society of the Promotion of Science
7. Committee on the International Prize for Biology (2004) The Twenty Years of International Prize for Biology. Japan Society of the Promotion of Science
8. Committee on the International Prize for Biology (2014) The Thirty Years of International Prize for Biology. Japan Society of the Promotion of Science
9. *Corner EJH (1990) His Majesty Emperor Hirohito of Japan. K.G. Biogr Memoirs Fellow R Soc 36:222–226
10. Crown Prince Naruhito (1993) Along with the Themes: Two Years, for Which I Studied in England (Themes to Tomoni – Eikoku no 2 Nenkan). Gakushuin Kyoyo Shinsho
11. Ebashi S (1999) The Royal Society of London awarded the King Charles II Medal to his Majesty the Emperor. Iden 53(2):6–7
12. Emori K (1998) His highness Prince Akishino (Akishino-no-Miya sama). Mainichi Newspapers
13. Emperor Akihito et al (1999) Manuscripts written by the Emperor. Bungeishunjyu, special issue of October
14. Habe T (1988) His Majesty the Emperor's researches in Biology. Umiushi Tsushin issued by Research Institute of Marine Invertebrates, no 1

---

[1]All references are written in Japanese except the articles with an asterisk. Japanese titles are shown only for the key references.

© Springer Nature Singapore Pte Ltd. 2019
H. Mohri, *Imperial Biologists*, Springer Biographies,
https://doi.org/10.1007/978-981-13-6756-4

15. Hamao M (1992) Crown Prince Hiro: record of fostering to be strong and sturdy over ten years. PHP (Peace and Happiness through Prosperity) Institute

16. Hara T (2008) Emperor Showa (Showa Tenno). Iwanami Shoten Publishers

17. Hatano M (1998) Report of Crown Prince Hirohito's visit to European Countries. Soshisha Publishing Co.

18. Hatano M (2011) Record of the attendance of Crown Prince Akihito at the coronation of Elizabeth II. Soshisha Publishing Co.

19. *Hutchinson GGE (1951) Marginalia. Am Nat 38:612–619

20. Imajima M (1999) Three-generation Imperial biologists—Researches of the Emperors, the Princes and the Princess. Kodomo no Kagaku (Natural Sciences for Children), vol 62, no 10

21. Imperial Household Agency (ed) (2015–2018+) The Annals of Emperor Showa (Showa Tenno Jitsuroku), vols 1–18+. Tokyoshoseki

22. Inoué S (2003) Acceptance address. In: Record of 2003 International Prize for Biology. Committee on the International Prize for Biology

23. Irie S (1965) Essays around the Imperial Moat (Horibata Zuihitsu). Bungeishunjyu Ltd.

24. Irie S (ed) (1981) Observers of Imperial Court 'School' (Kyu-chu Monzen Gakuha). TBS Britanica

25. Irie S (ed) (1985) Story of Court Chamberlains (Kyuchu Jijyu Monogatari). Kadokawa Publishing Co.

26. Irie T, supervised, Asahi Shimbun (ed) (1990–1991) Irie Sukemasa's Diary (Irie Sukemasa Nikki) 1-6. Asahi Shimbun

27. Isono N (1988) Those Who Came and Went from the Misaki Marine Biological Station—Birth of Zoology in Japan (Misaki Rinkai Jikkenjo wo Kyorai-shita Hitotachi – Nihon ni okeru Dobutsugaku no Tanjyo). Japan Scientific Societies Press

28. Ito M (2013) Botanical taxonomy. University of Tokyo Press

29. Ito Y (2011) A Life of Emperor Showa (Showa Tenno Den). Bungeishunju Ltd.

30. Kagaku Asahi (ed) (1991) A family of Noble Biologists (Tonosama Seibutsugaku no Keifu). Asahi Shimbun

31. Kageyama N (1999) Emperor Showa's self-actualization and his study of biology. The report of Tokyo University of Fisheries, vol 34, pp 25–41

32. Kakizawa R (2006) For a short tea time in the Institute, Princess Nori Sayako laughed so hard that tears came out. Zenjin, January number, p 21

33. Kanroji O (1957) The Emperor in a Suit (Sebiro no Ten'no). Tozai Bunmei Sha

34. Kasahara H (2001) Manual of Successive Emperors—How has the Imperial Throne been Succeeded? Chuko Shinsho

35. Kinoshita M (1998) New Version of Record of Experience in the Imperial Court—Serving Emperor Showa. Nippon Kyobunsya Co., Ltd.

36. Koyama T (1997) Story of the Botanical Garden. Aboc Press

37. Matsuura K (2011) Memories of Kazunori Takagi. Japan J Ichthyol 58:118–119

38. Maruyama K (2000) The View of the Place Unpromised—The Career of a Zoologist. Japan Scientific Societies Press

39. Mawatari S (1994) The Logic of Animal Taxonomy—Method of Recognizing Biodiversity. University of Tokyo Press

40. Mikuriya T, Iwai K (eds) (2007) Diary of Ryogo Urabe, the Last Chamberlain of Emperor Showa (Showa Tenno Saigo no Sokkin – Urabe Ryogo Jijyu Nikki), vols 1–5. Asahi Shimbun Publications Inc.

41. Minaka N (2006) The World of Phylogenetic Thinking—All along the Phylogenetic Tree. Kodansha Gendai Shinsho

42. Miwa N (1980) A consideration for the Emperor's visit to US in 1975. Moralogy Res 9:83–112

43. Mohri H (2010) Chasing My Dream Entrusted to Biology (Seibutsugaku no Yume wo Oimotomete). Minerva Publishing
44. Mohri H (2011) Memorandums of the Misaki Marine Biological Station, the University of Tokyo. Seibutsukenkyusha
45. Mohri H, Yasugi S (eds) (2007) The History of Zoology in Japan. Baifukan Co., Ltd.
46. Murata J, supervised, Ogura K (eds) (2004) Emperor Showa's Natural Museum (Illustrated book of Lake Hamana Expo). Shizuoka Expo'90 Foundation
47. Nagazumi T (1992) Serving to the Emperor for 80 Years—Emperor Showa and Me (80 Nenkan Osoba ni Tsukaete - Showa Tenno to Watakushi). Gakken
48. *Nakabo T (2011) *Oncorhynchus kawamurae* "Kunimasu", a deepwater trout, discovered in Lake Saiko, 70 years after extinction in the original habitat, Lake Tazawa, Japan. Ichthyol Res 58:180–183
49. Oba H (1996) History of Botanical Researches in Japan—the 300 Year History of the Koishikawa Botanical Garden. The University Museum, the University of Tokyo
50. Ochiai A (1987) Wandering about ichthyology. Kuroshio Journal of Kuroshio Science Institute, Kochi University, vol 2, pp 5–7
51. Ono K (1993) Certain Serious Incident in the Imperial Court. Kodansha
52. Oshima M (1940) The luckiest guy in Japan. Zool Mag 52(9):344–346
53. Our Imperial Family Editorial Department (2009) The Emperor's researches—co-worker talks "A side of the Emperor as the greatest taxonomist of gobies of the day". Our Imperial Family, no 45, pp 30–35
54. Prince Akishino Fumihito (2000) Fowls and Human Beings—From the Ethno-biological Viewpoint (Tori to Hito – Minzoku Seibutsugaku no Shiten kara). Shogakukan Inc.
55. Prince Hitachi Masahito (1999) Emperor Showa's research (Showa Tenno no Go-kenkyu). Our Imperial Family, no 2, pp 7–11
56. Princess Nori Sayako (2005) Days Went by—Anthology of Waka Poems and Messages by Princess Nori Sayako. Daito Publishing Co., Inc.
57. Research Group of Imperial Palace Flora and Fauna, the National Museum of Nature and Science (ed) (2001) Organisms Living in the Imperial Palace and Fukiage Garden. Sekaibunka com
58. Showa-Seitoku Memorial Foundation, the National Museum of Nature and Science (1988) His Majesty the Emperor's researches on biology. The National Museum of Nature and Science
59. Taki Y (2010) Imperial family and their biological Researches. J Tokyo Univ Mar Sci Technol 6:1–4
60. Tanaka T (1949) Emperor Showa and His Biological Research (Tenno to Seibutsugaku Kenkyu). Dainippon Yuben Kai Kodansha
61. Tei M (2011) Why Do the Emperors Study Biology? Kodansya
62. The National Museum of Nature and Science (ed) (2007) Fauna Sagamiana. Tokai University Press
63. The National Museum of Nature and Science (ed) (l988) The Emperor's Biological Research. The National Museum of Nature and Science
64. Tokyo University of Marine Science and Technology Library (2009) The Emperor's ichthyological researches. Exhibition Catalog
65. Tokugawa Y, Iwai K (1997) Grand Chamberlain's Will—Serving Emperor Showa for 50 Years (Jijyucho no Isho—Showa Tenno tono 50 Nen). Asahi Shimbun
66. Tomiyama I (1979) Materials collected in the University Museum, the University of Tokyo (11) Specimens of fishes. UP, no 75, pp 26–31
67. Ueno M (1987) Japanese History of Zoology. Yasaka Shobo

68. Watanabe M (2009) A Chamberlain of the Imperial Family—Serving as the Grand Chamberlain for ten and a half years (Tennoke no Shitsuji—Jijyucho no 10 Nen-han). Bungeishunju Ltd.
69. Watanabe M (2014) A Bridge between the Hearts Linked by the Her Majesty Empress Michiko. Chuokoron Shinsya, Inc.
70. Yamashina M (1966) Biologists, who were devoted to researches on hydrozoans— Masamaru Inaba, Eberhard Stechow, the Emperor Showa. Gakuto 93(4):16–19
71. Yamashina Y (1949) Classification of Animals Based on Cytology. Hoppo Shuppan Sha